Man and Environmental Processes

STUDIES IN PHYSICAL GEOGRAPHY
Edited by K. J. Gregory

Forthcoming volumes

Geomorphological Processes
E. DERBYSHIRE, K. J. GREGORY and J. R. HAILS

Ecology and Environmental Management
C. C. PARK

Applied Climatology
J. E. HOBBS

Other volumes are in preparation

STUDIES IN
PHYSICAL GEOGRAPHY

Man and Environmental Processes

A Physical Geography Perspective

Edited by

K. J. GREGORY
Professor of Geography, University of Southampton
D. E. WALLING
Reader in Physical Geography, University of Exeter

BUTTERWORTHS
London Boston
Durban Sydney Toronto Wellington

First published in Britain by Wm Dawson & Sons Ltd 1980
First Butterworths edition 1981

British Library Cataloguing in Publication Data

Man and environmental processes.—(Studies in physical geography ISSN 0142–6389).
  1. Geology   2. Man—Influence on nature
  I. Gregory, Kenneth John   II. Walling, Desmond Eric   III. Series
  551   QE625   80–41407

  ISBN 0-408-10736-7 Cased
        0-408-10740-5 Limp

Filmset in 10/12 point Times
Printed and bound in Great Britain
by Mackays of Chatham

Contents

Contributors

B. W. ATKINSON	Reader in Geography, Queen Mary College, University of London
E. C. F. BIRD	Reader in Geography, University of Melbourne
R. B. BRYAN	Professor of Geography, University of Toronto
D. R. COATES	Professor of Geology, State University of New York, Binghamton
H. M. FRENCH	Associate Professor of Geography, University of Ottawa
D. GREENLAND	Professor of Geography, University of Colorado and Institute of Arctic and Alpine Research, Boulder, Colorado
K. J. GREGORY	Professor of Geography, University of Southampton
C. M. HARRISON	Lecturer in Geography, University College London
W. R. ROUSE	Associate Professor of Geography, McMaster University
M. J. SELBY	Reader in Earth Sciences, University of Waikato
I. G. SIMMONS	Professor of Geography, University of Bristol
S. TRUDGILL	Lecturer in Geography, University of Sheffield
D. E. WALLING	Reader in Physical Geography, University of Exeter

Figures

Tables

1

Introduction

K. J. GREGORY and D. E. WALLING

The object of the present volume is: to indicate the character and, approximately, the extent of the changes produced by human action in the physical condition of the globe we inhabit; to point out the dangers of imprudence and the necessity of caution in all operations which, on a large scale, interfere with the spontaneous arrangements of the organic or the inorganic world; to suggest the possibility and the importance of the restoration of disturbed harmonies and the material improvement of waste and exhausted regions; and, incidentally , to illustrate the doctrine, that man is, in both kind and degree, a power of a higher order than any of the other forms of animated life, which, like him, are nourished at the table of bounteous nature.

G. P. Marsh, 1864

With these words George Perkins Marsh began the preface of his volume *Man and Nature* (1864). The book was conceived as a 'little volume showing that whereas others think that the earth made man, man in fact made the earth', it proved to have a great impact in the way in which men visualise and use the land, (Lowenthal, 1965) and became a fundamental basis for the conservation movement (Mumford, 1931). Although this significant book did include reference to the effects of human activity upon environmental processes its primary aim was to demonstrate the magnitude of the changes wrought by man. This is shown by the way in which chapters were devoted to vegetable and animal species (Chapter 2), to the woods (3), to the waters (4), to the sands (5), and to the projected or possible geographical changes by man (6). The aim of the present volume is to review the effects of human activity on physical environment processes, and this is justified not only as a complement to the approach taken by G. P. Marsh, but also as a sequel to the work produced since 1864, with contributions since the mid-nineteenth century to the study of the significance of human activity divided into three: those produced up to 1960 (Section 1.1), those produced between 1960 and 1970 (1.2), and those produced since 1970 (1.3).

1.1   A CENTURY OF MILESTONES

Although the importance of the study of human activity is now acknowledged as an integral part of physical geography, it is salutory that in the century after the publication of *Man and Nature* there were comparatively few major and comprehensive reviews by geographers of the subject afforded by this theme. Explanation of the paucity of contributions may be found in the fragmentation of physical geography and in the way in which progress in the subject was dominated conceptually by evolution. Whereas physical geography was viewed as an integral discipline in the nineteenth century, so that Huxley embraced air, land surface, vegetation, soil and organisms within the compass of his *Physiography* (1877), in the twentieth century physical geography increasingly separated into geomorphology, climatology, and biogeography. Emphasis on evolutionary concepts provided an intellectual environment of physical geography in which human activity was not accorded prime consideration. Evolutionary ideas which may have originated from the influence of Darwin's *Origin of Species* (1859) became incorporated in conceptual approaches throughout physical geography (Stoddart, 1966). The cycle of erosion came to dominate geomorphology, and parallels can be found in the concepts of plant succession, in the development of zonal soils, and in air mass climatology that pervaded other parts of physical geography.

Despite the evolutionary focus which dominated the branches of physical geography in the first half of the twentieth century there were several contributions which may be regarded as milestones in the elucidation of the impact of human activity upon physical landscape. One of these was provided by R. L. Sherlock in *Man as a Geological Agent* published in 1922. In this book, and in a related article (Sherlock, 1923), Sherlock developed the theme of the significance of man as an agent in geographical change. This was achieved by distinguishing geological and biological aspects. The latter, which included the effects on plant and animal species, was not included in his book which concentrated on geological aspects with particular attention devoted to denudation by excavation and attrition, to subsidence, to accumulation, to alterations of the sea coast, to the circulation of water, and to climate and scenery. Sherlock thus introduced the study of man, emphasized the contrasts between natural and human denudation, proposed that there are indications that the doctrine of uniformitarianism had been carried too far, and concluded that in a densely populated country like England, 'Man is many more times more powerful, as an agent of denudation, than all the atmospheric denuding forces combined' (Sherlock, 1922, p. 333).

Whereas this milestone was general in character, more specific ones became evident in the succeeding twenty years. These were inspired particularly by contemporary problems and although not devoted exclusively to the influence of human activity, books on soil erosion appeared reflecting growing awareness of a world-wide problem which necessitated conservation measures. Thus the book by G. V. Jacks and R. O. Whyte (1939) *The Rape of the Earth* provided a world survey of soil erosion as a basis for consideration of control measures, conservation and the political and social consequences. Perhaps most significant was an international symposium *Man's Role in Changing the Face of the Earth*, an interdisciplinary symposium with international participation which was organised by the Wenner-Gren Foundation for Anthropological Research and held at Princeton, New

Jersey in 1955. This meeting prompted an 1193-page volume with the same title (Thomas, 1956) which has subsequently been republished in parts. The intellectual stimulation for this impressive venture was acknowledged to be the work of G. P. Marsh and that of the Russian geographer, A. I. Woeikof (1842–1914) who had also developed, independently, a utilitarian approach to the study of the earth's surface which acknowledged the activities of man. This volume embraced more than fifty-two chapters collected in three parts. The first part was retrospective, elaborating the ways in which man has changed the face of the earth; the second reviewed many ways in which processes of seas and land waters, atmosphere, slope and soils, biotic communities, ecology of wastes, and urban and industrial demands upon land has been modified; and the third drew attention to the prospect raised by the limits of man and the earth and the role of man. This monumental achievement certainly provided a milestone in material available to earth scientists by collating research from a broad spectrum of earth science disciplines and by providing an exhaustive compendium of reviews of the state of knowledge in 1956. In the shadow of this large volume the present book must be conceived as a small contribution which focuses on process and particularly on the developments during the last two decades.

1.2   A DECADE OF PAPERS

Increasing attention accorded to the significance of human activity in the physical environment was subsequently apparent in general review papers and in more specific research contributions that were published in the decade beginning in 1960. A series of general review papers appeared and a sample of these reflect the general trends which were becoming established. In a view of *Man and the Natural Environment* Wilkinson (1963) illustrated the transmogrification of the face of the earth under the impact of the destructive, the conservative and the creative agency of man by illustrating the scale and extent of human activity in relation to the atmosphere, the lithosphere and hydrosphere, the soil, and the biosphere. The previous deficiency of studies of the form-creating activity of man and of the influence of man on natural phenomena was regretted by Fels (1965) who advocated the necessity of study of anthropogenous geomorphological processes. J. N. Jennings in a presidential address in Australia (Jennings, 1966) revived the title used by Sherlock, of 'Man as a Geological Agent', for his review in which he stressed that the greater significance accorded to man arises because of more studies of process and also because measurements of contemporary processes are nearly always heavily biased by anthropogenic effects.The geomorphological significance of man was also reviewed in a significant paper by Brown (1970) in which he characterised man as a geomorphological process in relation to his direct, purposeful modifications of landform and to indirect effects and this involved reference to the influence of human activity upon geomorphological processes. General statements also occurred in the USSR and for some this has involved the designation of the influence of man as generating a new geological epoch, the Noosphere. This concept involves the active role played by the conscious mind and science in the purposefully directed, and not simply haphazard, development of the man-nature relationship (Trusov, 1969). It is a concept which is capable of wider use in geography (Bird, 1963).

Perhaps most significantly at the end of this decade Chorley explored the role and relations of physical geography (Chorley, 1971) and this was succeeded by affirmation of the possibilities of a system approach (Chorley and Kennedy, 1971). One of the systems proposed was the control system which is concerned particularly with man and the way human activity operates as a regulator in natural systems. In the ecosystem state control merely involves the manipulation of the negative-feedback loops in order to stabilize the system operation at some optimum state (Chorley, 1973). In an evaluation of this approach in relation to others Chorley (1973) noted 'It is clear, however, that social man is, for better or worse, seizing control of his terrestrial environment and any geographical methodology which does not acknowledge this fact is doomed to inbuilt obsolescence'.

This sentiment was therefore emerging in a number of general conceptual papers published in the 1960s and the ideas have continued to develop subsequently as well. Thus the problem of Quaternary relief development and man was reviewed by Demek (1973) against the fact that 55 per cent of the world's dry land surface is intensively used by man, that 30 per cent is partly modified by man, and the remaining 15 per cent is only slightly modified if at all.

Such general movements in the focus of geographic enquiry were accompanied in this decade by at least four other trends within and beyond the canvas of writings by geographers. Developments within the subject included firstly an increasing awareness of the need to document and to study processes, and secondly growth of interest in the hazards presented by the natural environment pertinent to human activity. More widely across the spectrum of the earth sciences there was, thirdly, the initiation of international programmes which incorporated the effects of man, and fourthly, the emergence of world-wide concern for the exhaustibility of earth resources in relation to human activity on spaceship earth.

The first of these themes, the emphasis upon environmental processes, occurred as a necessary requirement to permit the further development of models of the development of the physical environment. Therefore as rates of erosion were deduced and compared for various world areas it was demonstrated, for example, that erosion rates near Rome, Italy are now between 100 and 1000m^3 km^{-2} per year whereas they were between 20 and 30m^3 km^{-2} per year prior to man's influence (Judson, 1968). A number of studies investigated the effects of man's influence expressed through land use pressure as reflected, for example, on the activity of gullying in the south-west USA (Denevan, 1967) and in fires in relation to floods in the Bow Valley of Alberta (Nelson and Byrne, 1966). In some areas study of processes, of the influence of man upon these processes, and of the environmental effects of man's role as a regulator of these processes, proceeded to introduce greater concern for an evaluation of the engineering strategies available to minimize man's influence. Thus in areas of permafrost greater knowledge of permafrost equilibrium in relation to human activity is a prelude to an understanding of the solutions whereby disturbance of the permafrost by man can be minimized (e.g. Ferrians, Kachadoorian and Greene, 1969).

In the field of climatology the effects of man upon atmospheric processes and in the creation of man-induced climates attracted studies including the analysis of urban climates epitomized by an examination of the climate of London (Chandler, 1965). Studies of man's effect on soils had been available for two decades in the form of works on soil erosion but these were complemented by investigations of the less obvious ways whereby soil processes and characteristics had been affected by man. Thus in the area adjacent to the Silesian industrial region, Gilewska (1964) identified the changes in the geographic environment brought about by industrialization and urbanization, including the incidence of metal ions in soils derived from emission into the atmosphere from industrial sources. The influence of man has always been acknowledged in the biogeographic field but this also received renewed attention, as for example by Fosberg (1963) who reviewed man's place in the island ecosytem. Greater use of the ecosystem concept thus allowed evaluation of the significance of man in particular situations such as the effects of catastrophic human interference with coral island ecosystems (Stoddart, 1966) and comparison of the productivity of agricultural with the natural systems which man replaces. A number of studies therefore embraced consideration of the significance of recent plant invasions (Harris, 1966).

A second theme, which often stemmed from consideration of process extremes, was introduced by investigations of natural hazards. Because a physical event becomes a hazard only when the mechanics of physical environment are treated in relation to human activity, this approach provided an additional impetus for the growing interest in the significance of human activity. The initial interest was devoted to the flood hazard but this was soon complemented by consideration of the range of environmental hazards (Burton, Kates and White, 1968). The way in which such an approach provided an integrated theme especially appropriate in geography is exemplified by the appearance of books like *The Value of the Weather* (Maunder, 1970) and *Water, Earth and Man* (Chorley, 1969).

Two additional influences which encompassed and affected, but did not originate within geography were provided by international programmes and by environmental concern. A number of international programmes were inaugurated and these included the International Hydrological Decade (1965–1974) which dealt with man's influence as one of the salient themes, and the later Man and the Biosphere programme (1970) which patently devolves upon the magnitude of human influence.

Environmental concern was generated during the decade and stemmed from salutory warnings such as *Silent Spring* (Carson, 1962), from consideration of *Man and Environment* (Arvill, 1967) and subsequently from debates about population and the limits to growth (Ehrlich and Ehrlich, 1970). Subsequently in a work entitled *The Environmental Revolution*, Nicholson (1972) provided a chart of human impacts on the countryside, of the areas affected and of the consequent effects.

1.3 A DECADE OF READINGS

One consequence of the 1960s may therefore be seen as the development of specific studies of man's influence on particular processes and sections of environment, accompanied by general conceptual attention accorded to human activity. These developments internal to geography were achieved within an intellectual environment which embraced growing concern for the effects of man in the past and concern for his future, and this provided one of the motivating reasons for the initiation of international research programmes.

The culmination of these trends and their continuation into the decade of the 1970s found the available geographic literature not entirely appropriate. This deficiency therefore was remedied by a number of books which were collections of papers or readings. Some of these were specially commissioned and the Association of American Geographers Commission on College Geography in 1971 conceived a collection of essays devoted to geographic research on environmental problems (Manners and Mikesell, 1974). Although the ensuing volume was not specifically devoted to human activity, this theme was prominent in each of its twelve chapters. An alternative variant in books produced in the 1970s was demonstrated by collections of previously published papers often derived from a wide spectrum of journals. Thus Detwyler (1971) collated previously published papers to provide *Man's Impact on Environment* and Coates (1972, 1973) provided volumes of readings on *Environmental Geomorphology and Landscape Conservation.* These volumes drew together papers from a wide range of environmental publications and also showed how papers of significance existed prior to 1970. In a volume devoted to non-urban areas (Coates, 1973) the papers were organized into sections dealing with man-induced terrain degradation, soil conservation, and landscape management, and of the twenty-six papers reprinted in the volume, half were originally published before 1960 indicating that a number of early significant papers existed even if they were sporadically disseminated throughout the literature.

Subsequently a number of volumes composed of edited collections or of readings appeared which were devoted to particular fields germane to the theme of man and environment. Most numerous amongst these were a number of volumes devoted to urban physical geography and this theme was covered broadly by *Urbanization and Environment* (Detwyler and Marcus, 1972) and more specifically by *Urban Geomorphology* (Coates, 1976).

1.4   A BOOK FOR PROCESSES

It was against this background of a shift in emphasis within physical geography which crystallized in the 1960s and which was supported in the 1970s by the publication of a number of edited books and volumes of readings, that this book was conceived. It seemed timely to envisage a volume specially written to review the significance of human activity to the range of environmental processes studied by the physical geographer. The processes considered are those which have traditionally been studied by the physical geographer, so that pollution and global element cycles are not included.

Such an endeavour cannot achieve completeness but the use of contributions from authors from several continents was designed to provide as international a perspective as possible. Although each author reviews the significance of human activity for physical landscape processes the approach adopted necessarily varies, partly because the emphasis always has and still does differ between the component branches of the subject. This volume is directed towards land system processes with chapters grouped under the major headings of atmosphere, hydrosphere, lithosphere, pedosphere, and biosphere, but with certain important ingredients of these spheres excluded, partly inevitably and partly as a reflection of the research activity of physical geographers and their tendency to ignore the oceans.

The Château de Val, reconstructed in the fifteenth century, is now virtually surounded by a reservoir created in 1951 when the dam was built at Bort-les-Orgues to impound the headwaters of the Dordogne for the generation of hydroelectric power. Today the Chateau figures within what is obviously a man-made lake and such direct influences of man are rapidly deduced. The less obvious and indirect effects of the man-impounded lake on the downstream flows and river channel, on the local water balance through evaporation, and on plant communities are less immediately apparent. It is therefore to such indirect effects on process that this book is dedicated. In *Man and Nature* (1864) George Perkins Marsh titled the final chapter 'Projected or Possible Geographical Changes by Man'. The range of future changes which could be induced by man is now much greater than could be imagined in 1864 but a knowledge of the possible consequences for landscape processes is just as significant! There are implications for physical geography because in the past we may, according to Hewitt and Hare (1973), have been captives of tradition. These authors argued that ' ... the geographer, when he analyses the properties of man–environment systems, must base himself on the central functions of that system, rather than on the traditional divisions of physical geography' (Hewitt and Hare, 1973). By focusing more on human influence on physical landscape processes we may achieve a vital and relevant physical geography.

REFERENCES

ARVILL, R., 1967, *Man and Environment* (Penguin, Harmondsworth).

BIRD, J. H., 1963, 'The noosphere: a concept possibly useful to geographers.' *Scot. Geogr. Mag.,* 79, pp 54–6.

BROWN, E. H., 1970, 'Man shapes the earth.' *Geogr. Jnl.,* 136, pp. 74–85.

BURTON, J., KATES, R. W., and WHITE, G.F., 1968, *The Human Ecology of Extreme Geophysical Events.* Natural Hazard Research Working Paper No. 1 (Department of Geography, University of Toronto).

CARSON, R., 1962, *Silent Spring* (Penguin, Harmondsworth).

CHANDLER, T. J., 1965, *The Climate of London* (Hutchinson, London).

CHORLEY, R. J., 1969 (ed.), *Water, Earth and Man* (Methuen, London).

1971, 'The role and relations of physical geography.' *Progress in Geography,* 3, pp. 87–109.

1973, 'Geography as human ecology', *Directions in Geography*, ed. R. J. Chorley (Methuen, London), pp. 155–69.

CHORLEY, R. J. and KENNEDY, B. A., 1971, *Physical Geography: A systems approach* (Prentice Hall, London).

COATES, D. R., 1972 (ed.) *Environmental Geomorphology and Landscape Conservation, Volume 1, Prior to 1900,* Benchmark paper in Geology (Dowden, Hutchinson and Ross, Stroudsburg).

1973 (ed.), *Environmental Geomorphology and Landscape Conservation, Volume III, Non-urban regions,* Benchmark Papers in Geology (Dowden, Hutchinson and Ross, Stroudsburg).

1976 (ed.), *Urban Geomorphology.* Geol. Soc. of America, Special Paper 174.

DEMEK, J., 1973, 'Quaternary relief development and man.' *Geoforum,* 15, pp. 68–71.

DENEVAN, W. M., 1967, 'Livestock numbers in nineteenth-century New Mexico and the problem of gullying in the Southwest.' *Ann. Assoc. Amer. Geog.,* 57, pp. 691–703.

DETWYLER, T. R., 1971, *Man's Impact on Environment* (McGraw-Hill, New York).

DETWYLER, T. R. and MARCUS, M. G., (eds.), 1972, *Urbanization and Environment: The Physical Geography of the City* (Duxbuy Press, Belmont, California).

EHRLICH, P. and EHRLICH, A. H., 1970, *Population, Resource, Environment: Issues in Human Ecology* (W. H. Freeman, San Francisco).

FELS, E., 1965, 'Nochmals: Antropogen Geomorphologie.' *Petermanns Geographische Mitteilungen,* 109, pp. 9–15.

FERRIANS, O. J., KACHADOORIAN, R. and GREENE, G. W., 1969, 'Permafrost and related engineering problems in Alaska.' *US Geol. Surv. Prof. Paper,* 678.

FOSBERG, P. R., 1963, *Man's Place in the Island Ecosytem,* Symposium Tenth Pacific Science Congress, Honolulu (Bishop Museum Press, Bishop).

GILEWSKA, S., 1964, 'Changes in the geographic environment brought about by industrialization and urbanization.' *Problems of Applied Geography,* 2, pp. 201–10.

HARRIS, D. R., 1966, 'Recent plant invasions in the arid and semi-arid south-west of the United States.' *Ann. Assoc. Amer. Geog.,* 56, pp. 408–22.

HEWITT, K. and HARE, F. K., 1973, 'Man and environment. Conceptual frameworks.' *A.A.G. Commission on College Geography Resource Paper No. 20.*

HUXLEY, T. H., 1877, *Physiography: an Introduction to the Study of Nature* (Macmillan, London).

JACKS, G. V. and WHYTE R. O., 1939, *The Rape of the Earth* (Faber, London).

JENNINGS, J. N., 1966, 'Man as a geological agent'. *Austral. Jnl. Sci.,* 28, pp.150–6.

JUDSON, S., 1968, 'Erosion rates near Rome, Italy.' *Science,* 160, pp. 1444–6.

LOWENTHAL, D. (ed.), 1965, *Man and Nature by George Perkins Marsh* (Harvard University Press, Cambridge, Mass.)

MANNERS, I. R., and MIKESELL, M. W., 1974, 'Perspectives on environment.' *Assoc. Amer. Geog.* Publication No. 13.

MARSH, G. P., 1864, *Man and Nature or Physical Geography as Modified by Human Action* (Charles Scribner, New York).

MAUNDER, W. H., 1970, *The Value of the Weather* (Methuen, London).

MUMFORD, L., 1931, *The Brown Decades: A study of the Arts in America* 1865–1895 (Dover, New York).

NELSON, J. G. and BYRNE, A. B., 1966, 'Man as an instrument of landscape change. Fires, floods and national parks in the Bow Valley, Alberta.' *Geog. Rev.,* 56, pp. 226–38.

NICHOLSON, M., 1972, *The Environmental Revolution* (Penguin, Harmondsworth).

SHERLOCK, R. L., 1922, *Man as a Geological Agent* (Witherby, London).

1923, 'The influence of man as an agent in geographical change.' *Geog. Jnl.,* 61, pp. 258–73.

STODDART, D. R., 1966, 'Darwin's impact on geography,' *Ann. Assoc. Amer. Geog.,* 56, pp. 683-98.

1966, 'Catastrophic human interference with coral island ecosystems.' *Geography, 53, pp. 25–40.*

THOMAS, W. L., 1956, *Man's Role in Changing the Face of the Earth* (Univ. of Chicago Press).

TRUSOV, YUILI, 1969, 'The concept of the noosphere.' *Soviet Geography,* 10, pp. 220–36.

WILKINSON, H., 1963, 'Man and the natural environment.' *Department of Geography University of Hull, Occasional Papers in Geography, No. 1.*

Atmosphere

2

The Radiation Balance

D. GREENLAND
University of Colorado and Institute of Arctic and Alpine Research, Boulder, Colorado

2.1   THE RADIATION ENVIRONMENT

It is no accident that an account of humans and the physical landscape should start with a consideration of radiation and the radiation balance. Radiation from the sun is the starting point for most physical systems on the earth's surface and in its atmosphere (Fig. 2.1). Radiation from the sun heats the atmosphere and the earth's surface. It is an energy source for moving the great currents of air around the globe. These, in turn, make the heat from radiation more equitably distributed over the earth so that a large part of the planet has temperatures in a range inhabitable for human beings. As an energy source for evaporating water, radiation drives the hydrological cycle of the earth thus providing, through precipitation, a continually renewed source of fresh water for the world's agriculture and life. Another vital aspect of radiation from the sun is its role as an energy input to ecosystems. Through the process called photosynthesis, solar energy is changed into the chemical energy of plants that are later consumed by humans and other animals to sustain their life.

Besides these phenomena that we think of as mainly acting on a large scale, radiation is also important at the scale of an individual. As you read this you are being continuously bombarded by flows of radiation from your surroundings – walls if you are inside, the sky, sun, trees and other surrounding objects, if you happen to be outside. At the same time your body is radiating away heat energy – which is one of the ways it prevents you from becoming too hot. If there were no other ways of heat energy transfer, your body temperature would be controlled by the difference between incoming and outgoing flows of radiation. This difference is called the radiation balance.

This chapter will first examine the nature of radiation and its flows and balances in the earth–atmosphere system. Then attention will be turned to the way that nature herself

FIG. 2.1 SUNRISE
The beginning of another cycle of shortwave radiation input to the Earth's physical system

alters the balance. Dust from erupting volcanoes, for example, has a marked effect. Most space, however will be devoted to the many ways in which humans can and do alter the radiation balance. This occurs on local, regional, and global scales, with interactions taking place between the scales. Considering the importance of radiation to life on earth, the implications of human influence on the radiation balance are sometimes staggering. Consequently, in conclusion it is appropriate to speculate on the effects of human influence on the radiation balance and the climate of the earth in the future.

2.2   RADIATION: ITS NATURE, FLOW, AND BALANCE

The question of what is radiation has puzzled scientists at least from the time of Sir Isaac Newton. There is still no absolute certainty, but there is a useful conceptual framework that satisfactorily explains most of the radiation-related phenomena that are observed. It is believed that a gas at absolute zero temperature (0 K or $-273\,°C$) would be at its lowest possible energy level. If the gas was heated, atoms would be raised to higher energy states and, at the same time, some would spontaneously drop back to lower energy levels. In such a drop, a 'bundle' of energy called a *photon* would be emitted. The photon is regarded as a particle of radiation that has the property of having a wavelength. Like several ocean waves coming into a beach the distance between successive crests is called the wavelength. Continuous emission of photons causes a continuous train of radiation of a certain wavelength or a situation analogous to sending waves across a pond by disturbing one side of it at regular intervals.

There are two laws basic to the understanding of the flow and nature of radiation. The first, known as Wien's displacement law, states that the wavelength of the radiation is inversely proportional to the temperature of the emitting body. Thus the sun radiating at 6000 K has radiation with wavelengths of between 0·2 to 10·0 microns. (1 micron, μm, is 1 millionth of a metre). This is called shortwave radiation in the meteorological context, and is short relative to the longwave radiation from the earth's surface and atmosphere. The latter radiate at about 300 K and the consequent radiation has wavelengths between 4 and 50μm. A second law is the Stefan-Boltzmann law which tells us that the amount of flow, or flux, of radiation is proportional to the fourth power of the absolute temperature of the radiating body. Therefore, judging by the temperatures given above, it is to be expected that a far greater amount of radiant energy is emitted from the sun than from the earth.

The results of the action of these two laws are seen in Figure 2.2. The dashed lines show the amount of radiation coming from perfect emitters at 6000 K (left) and 300 K (right). The solid lines indicate, in the short wavelengths, the amount of direct-beam radiation arriving at the earth's surface, and, in the long wavelengths, the value of longwave radiation lost to space from the earth's surface. The irregularity of the solid lines indicates the many wavebands where both short and long wave radiation are absorbed by atmospheric constituents. One of the most prolific absorbers is ozone which accounts for the absence of radiation of less than 0·4μm.

Whereas the physical laws are essential to understand the nature and intensity of radiant flow in the atmosphere, it is more practical to describe how the various flows of radiation to and from the earth's surface may be measured. Figure 2.3 illustrates the necessary measuring sensors. In this case they are located at 3490 m near the continental divide in the Rocky Mountains USA. The two glass domed sensors on the left measure incoming short wave radiation. They measure both the *direct radiation* from the sun (S) and *diffuse radiation* (D) that arrives at the earth's surface after being scattered from molecules of the atmosphere and clouds. The direct and diffuse radiation together make up *incoming shortwave radiation* ($K\downarrow$). These two sensors are not identical since one has a glass dome which lets through the visible part (0·40–0·74μm) of short wave radiation, and the other has a quartz dome permitting the passage of both visible and ultra violet (0·15–0·4μm) light. Thus by subtracting the recorded values of the latter from the former, it is possible to estimate the receipt of ultra violet light. The shortwave sensor in the centre of the platform is surrounded by a shadow band which prevents direct shortwave radiation. Some of the shortwave (and in fact longwave) radiation that reaches the ground is reflected. Two sensors in the centre at the right-hand end of the platform measure incoming shortwave ($K\downarrow$) and *reflected shortwave radiation* ($K\uparrow$). The ratio of these two ($K\uparrow/K\downarrow$) is called the *albedo* (α) of the surface. The albedo is a critical part of the radiation balance when it comes to the alteration of the radiant flows. Finally, three of the sensors at the right-hand end of the platform measure various flows of long and shortwave radiation together. Since both the earth's surface and the atmosphere have temperatures

D. GREENLAND

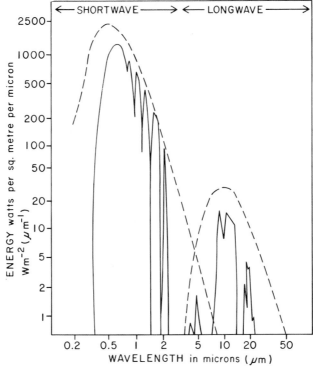

FIG. 2.2 DISTRIBUTION OF RADIATION INTENSITY WITH WAVELENGTH FOR ATMOSPHERIC RADIATION
(Redrawn from *Physical Climatology* by W. D. Sellers, by permission of University of Chicago Press)

FIG. 2.3 PLATFORM WITH SENSORS FOR MEASURING VARIOUS FLOWS OF RADIANT ENERGY TO AND FROM THE
EARTH'S SURFACE

above absolute zero (0 K) they both radiate *longwave radiation* (*L*). The operative sensors on these three instruments are covered by polyethylene which permits the passage of both long and short wave radiation. One instrument measures net radiation (Q^*), the algebraic sum of the total upward and downward short and longwave flows. Another measures the upward fluxes ($K\uparrow + L\uparrow$) while the third monitors the downward flow ($K\downarrow + L\downarrow$). Subtraction of the measured flows of $K\downarrow$ and $K\uparrow$ leads to an estimate of the upward ($L\uparrow$) and downward ($L\downarrow$) flow of longwave radiation.

The actual measurement of the flows of radiation to and from the earth's surface, and within the atmosphere is therefore quite complex. However, calculations and checks on the measurements are made easier by the fact that there is a radiation balance. This is in fact a special form of the law of conservation of energy and says that the balance of radiation at the earth's surface is the sum of upward and downward flows of short and longwave radiation. The concept can be applied not only to the earth's surface but to higher levels of the earth–atmosphere system as a whole. Using the symbols defined above, the radiation balance or net radiation can be expressed in several forms:

$$Q^* = K\downarrow - K\uparrow + L\downarrow - L\uparrow$$
$$Q^* = K\downarrow(1-\alpha) + L\downarrow - L\uparrow$$
$$Q^* = (S+D)(1-\alpha) - (L\uparrow - L\downarrow)$$

Three terms are very important. First the incoming shortwave flow $K\downarrow$, secondly the albedo, and thirdly the longwave balance between $L\uparrow$ and $L\downarrow$. In practice the shortwave and longwave flows are markedly affected by the *turbidity*, that is the degree of impurity of the atmosphere. The albedo is determined largely by the nature of the earth's surface or the content of the atmosphere at a particular location. The longwave radiation balance is mainly controlled by the temperatures of the surface and the atmosphere and these depend in turn partly on the nature and composition of the surface and atmosphere respectively. The situation will become clearer upon examination of some actual examples.

2.3 NATURAL ALTERATION OF THE RADIATION BALANCE

There are many variations in the composition of the atmosphere due to changing humidity and cloud conditions and the variation of impurity content. It is likely that no-one would like to define the point at which atmospheric composition actually becomes unusual. However, there are some states which are clearly abnormal. One example is that due to excessive volcanic activity. Atmospheric composition will effect the quantity of radiation reaching the surface. This quantity could also be affected if the output of the sun changed. Solar variability will be examined first and then we will return to volcanoes.

The amount of solar radiation of all wavelengths received per unit time and unit area at the top of the earth's atmosphere is called the *solar constant*. Its most probable value is 1373 ± 20 W m^{-2} (Eddy 1977, p. 22). Note that present technology can measure this value only within limits of ± 1.5 per cent. What would be the effect of a change of this output on the earth's climate? Again the state of our present knowledge does not provide a complete answer. Schneider (1977, p. 18) has examined the sensitivity of the global surface temperature to changes in the solar constant. He concludes that the changes of global surface temperature of 1 K that have occurred in the last few thousand years could be explained by a change in the solar constant of somewhere between 0·2 and 2 per cent. These values are either within or near present measurement accuracy. Is it likely that such changes in incoming shortwave radiation have occurred? Recent analyses of historical and modern evidence of solar output suggest an affirmative answer. Such analyses show that in the last 1000 to 2000 years solar activity passed through three or four major departures from the present-day, rather benign, 'normal'. Solar activity matches climatic changes on earth quite well. The activity was high during the warm period in the eleventh to thirteenth centuries and showed two lows during the late fifteenth to early eighteenth centuries during the 'Little Ice Age'. Since about 1715 solar activity has generally been climbing and the increase has been paralleled by a general warming of world climate (Eddy 1977, p. 69). Thus, it must be concluded that any changes in the transparency of the atmosphere and changes in the radiation balance due to human activity must be viewed against a background of a solar constant that is quite possibly not constant.

In addition to the possibility of changing solar output, the transparency of the atmosphere may alter, especially when large amounts of dust are put into it. Volcanic eruptions spew enormous masses of particulate matter and gases into the atmosphere. The Krakatoa Eruption of 1883 is said to have injected 54 Km³ of ash, lava, and mud into the atmosphere. Such an event will obviously alter the transparency of the atmosphere and the radiation balance. Following this eruption, shortwave radiation arriving at the earth's surface was reduced by about 10 per cent. Reductions up to 25 per cent were recorded at some locations. Bryson and Murray (1977, p. 147) quote a report that following the eruption the Fire Department in Poughkeepsie, New York were called out because the brilliant sunset, due to the dust in the atmosphere, was mistaken for the glow of a fire in the western part of the city. The decrease of shortwave radiation was documented thoroughly following the eruption of Mount Agung on the Isle of Bali in 1963 (Ellis and Pueschel, 1971).

The injection of volcanic dust into the atmosphere decreases the direct shortwave radiation but can increase the diffuse shortwave radiation. The latter explains the spectacular sunsets following an eruption. Stringer (1972, p. 233) suggests that in general the direct beam is reduced by 20 per cent but, because of the increase in diffuse radiation, the net decrease in total radiation is just below 10 per cent. Volcanic dust in the atmosphere tends to cut out radiation reaching the earth's surface, but the particles concerned are mainly small and cannot hold much heat. Lamb (1970, p. 461) has called this a 'reverse greenhouse effect' and has suggested that it can lead to periods of decrease of earth temperatures after times of major volcanic activity.

There seems little doubt that volcanic dust can affect the radiation balance and the world's climate in the short term. There is some evidence that effects can be felt on a longer time scale. Lamb (1972, p. 433) quotes Fuch's evidence of three major long periods of extreme volcanic activity in Africa during the lower, middle and upper Pleistocene periods. These are suggested to coincide with three major periods of increased precipitation in East Africa. However, at the present time, the effect of volcanic activity on this and longer time scales must remain in a stage of speculation.

2.4 THE HUMAN INFLUENCE ON THE RADIATION BALANCE

Changes of solar output and volcanic dust are just two of a myriad of natural physical factors that can alter the radiation balance. Set against such a background is the relatively new presence of human influence on the balance. This influence also takes many diverse forms – some of which are certain and some of which also remain in the realm of speculation. Because of its diversity, human influence on the radiation balance is, perhaps, most appropriately illustrated with reference to different spatial scales. Whilst clearest at the local scale, the effects of human activity at the global scale still have to be brought more sharply into focus.

2.4a The local scale

No surface on the earth is artificially altered more than the urban surface. The activities on this surface also have a marked effect on the air above the surface (Fig. 2.4). Thus both the alterations to the surface and the overlying air affect the radiation balance in a significant way. The experience of two cities illustrate this point.

Measurements in Budapest, Hungary show that the annual mean decrease in incoming shortwave radiation near the city centre compared to the outskirts is about 7–8 per cent. This decrease, due to the polluted air over the city, rises to more than 15 per cent in the winter months from November to March (Probald, 1972). The Budapest study indicates that part of this loss of radiation is redressed by the change in albedo of the urban

FIG. 2.4 POLLUTION DOME OVER DENVER, COLORADO

surface. With the exception of two months in the year, the city always has a lower albedo (thus reflecting less radiation) than the rural surroundings. The absorbed shortwave radiation warms the urban atmosphere and consequently the downward longwave radiation is increased. Although the longwave radiation emitted from the city surface is also rather higher than that from rural surfaces (because of the greater surface temperatures), the difference is offset by the incoming longwave radiation. This happens in Budapest despite the fact that the incoming longwave radiation is only increased by about 1 per cent over that received in the city outskirts. One might expect this effect to be more marked in an area of concentrated heavy industry.

Such an area was investigated by Rouse and McCutcheon (1972) who instrumented sites near a steel mill and on the outskirts of Hamilton, Ontario. They found a similar attenuation of incoming shortwave radiation in the industrially effected atmosphere. The decrease averaged 9 per cent over all the records collected (mostly in summer) and rose to a maximum of 17 per cent. These investigations were able to detect the increased longwave downward radiation near the industrial site and this increase amounted to an average of 11 per cent with by far the greatest increase occuring during the day. Since the increase is not noted during the night, it is concluded that photochemical absorption of sunlight heats the industrial atmosphere and in turn the downward longwave radiation is amplified. It is interesting to note, however, that the changes in the various radiant flows tend to balance each other out, leaving the net radiation of the urban surface rather similar to that of the surrounding surfaces. The effect of increased longwave radiation countering decreased shortwave radiation was seen in both Budapest and Hamilton.

The modified urban atmosphere represents an inadvertant modification to the radiation balance. In contrast it should be mentioned that there are many practical ways of altering the radiation balance with beneficial effects. Perhaps the best-known of these is in the construction of a greenhouse. Here the glass permits the passage of incoming shortwave radiation but prohibits the passage of outgoing longwave radiation thus providing an overall heating effect. The same effect occuring on different scales in the atmosphere is, in fact, called the 'greenhouse effect'. The alteration of albedo on a small scale is also deliberately undertaken in some places. Carbon dust or dark material, for example, is sometimes spread on snow surfaces to promote the absorption of incoming radiation and the consequent melting of the snow. Some scientists believe the use of carbon black has potential on larger scales, as we see in the beginning of the next section.

2.4b   The regional scale

Most deliberate weather modification operations in the past three decades have been concerned with cloud seeding for the purposes of precipitation enhancement (cf. Chapter 3). In the future it may be that other areas are explored. One such area intimately related to the radiation balance is the use of carbon-black dust.

It is known that carbon aerosol particles of $0.1 \mu m$ or less maximize solar absorption per unit mass. Workers at Colorado State University hypothesize that significant beneficial influences can be achieved by using this absorption potential (Gray et al, 1974). The absorptance is so great that values of 1.8×10^{10} joules.kg^{-1} hr^{-1} or 3.3×10^{11} joules.hr.$^{-1}$ pound sterling^{-1} are believed to be achievable. Existing observational and modelling information suggests that this artificial heat source can be best used on a regional scale of about 100-200 km for several purposes. These include precipitation enhancement, the alleviation of tropical storm destruction, and to increase the rapidity of fog burn-off and snow melt.

As an example of the kind of procedure envisaged, the case of the tropical storm will be briefly examined. The assumption here is that if the outer boundary layer of a hurricane can be artificially warmed at about 0·5 to 1·0°C hr⁻¹ for ten hours, the vertical movement in the outer parts of the storm will be increased enough to reduce the low-level inflow to the centre of the system. This would lead to a decrease in storm damage. Gray and his associates suggest dropping carbon black from 10–20 jumbo-type aircraft concentrating on the right semi-circle of the system and outside the cirrus shield of the storm. The carbon black dust particles heat the surrounding air, primarily by direct shortwave absorption and rapid molecular conduction of the heat to surrounding air. Immediate longwave radiation transfers to the surrounding air are small. The air is then subject to enhanced convectional activity which eventually produces the desired effect.

Such schemes may seem fanciful and certainly cannot be considered reliable until thoroughly tested in the real world. However they do indicate the possibility of altering the radiation balance for beneficial purposes on a regional scale. There is growing evidence that the radiation balance can be altered by other factors at this scale. The case of changing the land surface for agriculture shows benefits in some cases, and drawbacks in others. The case of extensive urbanization presents likely deficits to human wellbeing.

It is estimated that 18 to 20 per cent of the land surface of the globe has been influenced by humans – primarily for agricultural purposes (SMIC, 1971, p. 63). All of these changes alter the radiation balance in some manner. For example a representative albedo for arable land or grassland is 20 per cent, which will mean less radiant heat being absorbed compared to coniferous forest with an albedo of 12 per cent or deciduous forest whose albedo might be 18 per cent (SMIC, 1971, p. 173). Alternatively, the irrigation of land in dry areas often gives rise to a green surface with an albedo of some 5 per cent less than a surrounding desert area (Fig.2.5). Most cases of land areas changed for food production are obviously beneficial to mankind. In rare, but important exceptions the reverse may be true.

An example of the latter may be the Rajputana desert in western India. Bryson and Murray (1977, p. 113) suggest that over-intensive use of the land in this area resulted in a change from a relatively fertile land system to a self-perpetuating desert. They have shown that the amount of water present in the atmosphere over this desert is quite comparable to that over some tropical forest regions, and suggest that dust in the atmosphere, caused initially by over-use of the land, now causes increased atmospheric subsidence inhibiting precipitation. The effect of this dust on the radiation balance according to Bryson and Murray (1977, p. 112) is threefold. First, there is a reflection of incoming radiation leading to lower surface temperatures than otherwise would occur, thus retarding the development of instability and rainfall production. Secondly, radiation in the long wavelengths at night from the top of the dust cloud results in cooling and subsidence of air. Thirdly, the trapping of longwave radiation from the ground at night can give temperatures high enough to prevent dew formation. The fact that the Rajputana

FIG. 2.5 CONTRAST IN SURFACE COVER BETWEEN AN IRRIGATED AREA AND SURROUNDING DESERT
Sevier Valley, Utah

desert is caused by human influence is disputed by the evidence of Singh *et al.* (1974), but the example shows quite plausibly a possible interrelation between human activity and the radiation balance.

Whether or not the Rajputana desert has been produced by human activity, there is no doubt that the development of large agglomerated urban areas can affect the radiation balance on a regional scale. Satellite photographs show the drift of haze from the eastern United States to the central Atlantic. Flowers *et al.* (1969) have demonstrated that turbidity values in the populated areas of the eastern US can be twice those of the relatively sparsely populated west. The impurity of air on this scale is likely to have similar effects as we found in the case of the individual city air. In addition it is suggested (SMIC, 1971, p. 217) that an increase in particles characteristic of heavy industrial pollution may decrease the albedo over large areas of land surfaces. However, little work has been done on this topic at this scale and there remains a pressing need.

2.4c   The global scale

The increasing urbanization of the planet is but one factor that might have effects on the radiation balance that cross the regional scale to the global scale. Many of the other factors already mentioned when considered en masse may well also affect the global radiation balance. As with the other scales, the global radiation balance might most easily be altered by either changing the albedo of the planet or by modifying the atmosphere's internal radiative processes. Examples of the former might be achieved by altering the extent of areas covered by snow, ice, or cloud. Similar effects result from changing the reflectivity of clouds, adding particles to the atmosphere, and excessive additions of water vapour by high flying aircraft. Examples of the latter include the consequences of small changes in the concentrations of some minor gases such as carbon dioxide and ozone.

Several of these changes involve quite spectacular events such as the Russian suggestion for melting the Arctic Ice by pumping warm Pacific water across a dammed Bering Strait, or the possible increase in the Earth's surface temperature of 0·5°C by the end of the century (assuming an increase of carbon dioxide without any negative feedback effects).

Many implications accompany such possibilities for global change. Some of these are illustrated by the recent debate on the possible global change of the ozone concentration in the stratosphere. Ozone protects life on earth from lethal doses of ultra-violet radiation. The effectiveness of the gas is demonstrated in Figure 2.2 by the sharp cut-off of radiation in the ultra-violet wavelengths reaching the Earth's surface. A decrease in the total ozone by 5 per cent could lead, at the very least, to an increase of nonmelanoma skin cancer (Smith et al., 1973). It is accepted that stratospheric ozone could be affected by nitrogen oxides emitted by high-flying aircraft (Johnston, 1971) and it may be that the same gases are finding their way into the stratosphere from the surface (McElroy et al., 1976). Another possible danger comes from the chlorocarbon industry. Crutzen et al (1978) calculated that emissions previous to 1976 have already reduced the global total ozone content by 1·5 per cent but uncertainties in knowledge of stratospheric chemistry cause such estimates to be treated with some reservations.

The implications of this example lead us to consider the future for the Earth's radiation balance.

2.5   THE FUTURE

The first implication is that major changes to the radiation balance resulting from human influence are quite possible. Secondly, the changes associated with the ozone example, as in many that appear above, are detrimental to human (and other) life on earth. Thirdly, despite considerable research there is still doubt surrounding the magnitudes and rates of change involved. Fourthly, there is even greater uncertainty as to the resulting effects on other parts of the radiation balance and the earth-atmosphere system. All of these implications hold good for the majority of radiation-balance changes that humans have effected on the earth whether the alteration be at a very local or a broad goegraphical scale.

The radiation balance is the key to the climate of the planet and any alteration of the balance should be approached with extreme caution for fear of producing irreversible effects. Many people recommend the use of models to make predictions of future changes in the radiation balance and other factors. But if rates of change are as fast as those quoted for removal of stratospheric ozone then there may not be time to construct adequate models. It is thus imperative that the development of new models of the radiation flow through the atmosphere, the changing radiation balance and its linkages with other aspects of the physical world, should be accompanied by a decrease in the rates of production of potentially harmful industrial products.

In 1957, Sir Fred Hoyle wrote a captivating novel called *The Black Cloud*. In it he described in fictional terms what would become of the earth's climate if a black cloud were interposed between the sun and the earth (Hoyle, 1957). This is science fiction at its best. However, it is a sobering thought that it might well be within the power of human beings to alter the radiation balance of the planet to an extent that equals or even supersedes natural alterations.

REFERENCES

BRYSON, R. and Murray, T. J., 1977, *Climates of Hunger* (University of Wisconsin Press, Madison, Wisconsin).

CRUTZEN, P. J., ISAKSEN, I. S. A. and McAFEE, J. R., 1978, 'The impact of the chlorocarbon industry on the ozone layer.' *Jnl. Geophys. Res.,* 83, pp. 345-63.

EDDY, J. A., 1977, 'Historical evidence for the existence of the solar cycle', *The Solar Output and its Variation,* ed. O. R. White (Colorado Associated University Press, Boulder, Colorado).

ELLIS, H. T. and PUESCHEL, R. F., 1971, 'Solar radiation: absence of air pollution trends at Mauna Loa.' *Science,* 148, pp. 493-4.

FLOWERS, E. C., McCORMICK, R. A. and KURFIS, K. R., 1969, 'Atmospheric turbidity over the United States 1961-6.' *Jnl. Appl. Met.,* 8, pp. 955-62.

GRAY, W. M., FRANK, W. M., CORRIN, M. L., and STOKES, C. A., 1974, *Weather Modification by Carbon Dust Absorption of Solar Energy.* Atmospheric Science Paper No. 225, Department of Atmospheric Science (Colorado State University, Fort Collins).

HOYLE, F., 1957, *The Black Cloud* (Harper and Row, New York).

JOHNSTON, H. S., 1971, 'Reduction of stratospheric ozone by nitric oxide catalysts from supersonic transport exhausts.' *Science,* 173, pp. 517-22.

LAMB, H. H., 1970, 'Volcanic dust in the atmosphere: with a chronology and assessment of its meteorological significance.' *Phil. Trans. of the Roy. Soc. of London: Mathematical and Physical Sciences,* 266 (1178), pp. 425-533.

1972, *Climate: Present, Past and Future. Vol. I Fundamentals and Climate Now* (Methuen, London).

McELROY, M. B. J., ELKINS, J. W., WOFSKY, S. C. and YUNG, Y. L., 1976, 'Sources and sinks of atmospheric N_2O.' *Rev. of Geophys, Space Phys.,* 14, pp. 143-50.

PROBALD, F., 1972, 'Deviations in the heat balance: the basis of Budapest's urban climate', *International Geography,* 1972, 1, ed. W. Peter Adams and F. M. Helleiner (University of Toronto Press. Toronto), pp. 184-6.

ROUSE, W. R. and McCUTCHEON, J. G., 1972, 'The diurnal behaviour of incoming solar and infra-red radiation in Hamilton, Canada.' *International Geography,* 1972, 1, ed. W. Peter Adams and F. M. Helleiner (University of Toronto Press, Toronto), pp. 191-6.

SCHNEIDER, S., 1977, 'Solar variability: a summary — the troposphere', *The Solar Output and its Variation.* ed. O. R. White (Colorado Associated University Press, Boulder, Colorado).

SINGH, G., JOSHI, R. D., CHOPRA, S. K. and SINGH, A. B., 1974, 'Late quaternary history of vegetation and climate of the Rajasthan desert, India.' *Phil. Trans. of the Roy. Soc,* B 267, pp. 467-501.

SMIC: Report of the Study of Man's Impact on Climate, 1971, *Inadvertant Climate Modification* (Massachusetts Institute of Technology, Cambridge, Mass.).

SMITH, K. C., BENER, P., Caldwell, M. M., DANIELS, F., GIESE, A. C., GOLDSMITH, T. H., KLEIN, W. H., LEE, J. A. H. and URBACK, F., 1973, *Biological Impacts of Increased Intensities of Solar Ultra Violet Radiation* (National Academy of Science, Washington, D.C.), pp. 1-46.

STRINGER, E. T., 1972, *Techniques of Climatology* (Freeman and Co, San Francisco).

3

Precipitation

B. W. ATKINSON
Queen Mary College, University of London

3.1 INTRODUCTION

Within the last two decades the atmosphere has increasingly been viewed as a resource (Chandler, 1970) and the relationships between man's activities and atmospheric behaviour have received renewed prominence in national scientific and technological programmes. Geographers may be tempted to argue that we are merely witnessing a delayed appreciation of some facets of climatic determinism! But sectarian interests and methodological squabbles apart, all scientists agree that in an increasingly crowded world, it is prudent, if not yet necessary, to reappraise the relationships between climate and man. Of the climatic elements, temperature and precipitation, particularly the former, have received the most attention as indicators of climatic change. The main reason for this emphasis is the availability of data. From the point of view of life on the planet, precipitation is arguably more important than temperature, as the droughts in the Sahel in the early 1970s and in the UK in 1976 graphically illustrated. The populations suffering those droughts were the latest in a historical succession where man has been essentially at the mercy of Nature's whims. Consequently any suggestion that man may be able to influence the amounts and distribution of precipitation is eagerly examined. The results of this examination over a period of thirty years or so fall conveniently into two classes: first, man's conscious attempts to influence precipitation; and secondly, his inadvertent effects upon precipitation. This chapter covers these two topics.

3.2 PRECIPITATION MECHANISMS

Before considering man's effects upon precipitation in detail, a brief review of precipitation types and mechanisms may be of value. The main precipitation forms are rain, snow and hail. All these forms of precipitation ultimately owe their existence to the cooling of air, usually by adiabatic expansion in uplift, so that the water vapour condenses or turns into ice. But this apparently simple process involves a myriad of mechanisms which are far from fully understood despite over two decades of intensive research (Mason, 1971). We can however paint a broad picture.

The condensation of vapour and freezing of water that may result from uplift and cooling require, respectively, condensation and ice nuclei. The resultant clouds may comprise any or all of the following: water droplets with temperatures greater than 0°C, supercooled water droplets with temperatures less than 0°C, and ice crystals. A major problem of cloud physics was the explanation of the growth of these cloud particles to a size which would cause them to fall out of the cloud as precipitation. Two mechanisms were suggested over thirty years ago and, in essence, are still accepted.

The more obvious of the two mechanisms is that involving coalescence and growth of particles. The term 'coalescence' in fact covers three types of 'joining together'; first, in a more restricted sense, coalescence means the joining of two liquid water drops; secondly, if ice crystals collide and join they are said to 'aggregate'; and thirdly, if an ice crystal collects a water drop the process is known as 'accretion'. Whichever of these processes operates, the resultant particle will be larger and, as such, its ability to 'catch' more cloud droplets or ice crystals dramatically increases. Of the three main types of precipitation, rainfall results largely from coalescence, snowfall from aggregation and hailfall from accretion.

The second, and far less obvious, way in which cloud particles may grow to precipitation size is known as the Bergeron or ice-crystal mechanism. This process depends upon the fact that saturated vapour pressure is greater over liquid than over ice surfaces. Consequently, if ice-crystals and supercooled water co-exist in a cloud (as they frequently do in extra-tropical latitudes), then the liquid particles tend to evaporate and the ice crystals to grow. Since ice particles are relatively few in number, each one draws on a comparatively large water supply and may grow into a crystal large enough to have an appreciable fall speed. The process of aggregation then proceeds rapidly, resulting in snowflakes that collect more cloud droplets and fall to lower, warmer altitudes where they may melt and continue descent as raindrops.

Both mechanisms outlined above are on the micro-scale, and can operate successfully only if cloud size, duration and internal airflow are favourable. Indeed the type, amount and intensity of precipitation from any cloud are largely determined by cloud dynamics. In the broadest terms, clouds must be about 1 km deep, last for about one hour and maintain suitable uplift (ranging from a few cm s^{-1} to several m s^{-1} depending upon cloud type) for any precipitation to occur. Stratiform clouds generally produce light, steady, quite prolonged rainfall or snowfall (largely dependent upon season) whereas cumuliform clouds may produce showers of rain or hail, often quite heavy in summer.

With this background knowledge of the micro- and macro-physics of clouds and precipitation we are in a position to review the ways in which man may influence them.

3.3   CONSCIOUS MODIFICATION OF PRECIPITATION

Man's conscious attempts to modify precipitation are of two main types: to increase amounts of rain and snow for agricultural and water-supply purposes; and to decrease the amounts of hail which cause so much damage in many agricultural areas. In both cases the modification is concerned with both the macro- and micro-physics of the clouds, particularly the latter.

3.3a   Increasing rainfall and snowfall

As a result of thirty years of effort stimulated by Schaefer's results (1946), many, but by no means all, meteorologists believe that 'cloud seeding' can indeed increase precipitation. Cloud seeding works in two ways. First, artificial nuclei may stimulate small cloud particles to coalesce; and secondly, cloud seeding with freezing nuclei or solid carbon dioxide (dry ice) may induce freezing and trigger the Bergeron process. In the first case, known as warm seeding, water drops may be introduced into a cloud to start a sweep-out process which otherwise is expected to be too long delayed. Because a tremendous weight of water is required to modify a whole cloud with large drops, finely divided salt or a hygroscopic liquid mist is usually used instead. For example, experiments by the Oklahoma Bureau of Reclamation used a concentrated water solution of ammonium nitrate and urea, sprayed from an aircraft into a cloud in the form of droplets about 0·02 mm in diameter. Within a minute the nitrate and urea droplets grew by gathering condensation from the vapour to a 0·05 mm size, a factor of 15 greater in mass. The 0·05 mm drops were large enough to start a sweep-out process that may have produced 5 mm diameter drops only 20 minutes later (Kessler, 1973).

The second method, seeding by dry ice or silver iodide, produces the same effects by two different processes. If dry ice is used it has the effect of inducing a massive rapid cooling which freezes the supercooled water. In contrast, silver iodide particles are good nuclei for ice formation because of the close resemblance of their crystal structure to that of ice. The particles do, in fact, act as freezing nuclei (Vonnegut, 1949). Whichever seeding material is used, the result is the production of ice crystals, which, it is argued, will trigger the Bergeron mechanism.

Hygroscopic particles and freezing nuclei can be admitted to the atmosphere in different ways. In the first field experiments, dry ice was dispersed from a small aeroplane (Schaefer, 1946) and silver iodide was generated at the ground (Vonnegut, 1950). This latter process involved mixing the iodide with acetone and water and spraying the solution into a hydrogen flame. The resultant smoke of potential freezing nuclei was then carried up into a cloud by natural convection currents. More recently silver iodide has been released from aircraft and pyrotechnic devices, with the aim of placing the nuclei more accurately in potentially profitable parts of a cloud. The whole exercise is still rather bedevilled by the finding that sunlight can reduce the power of the silver iodide to act as a nucleant.

Notwithstanding the above difficulties, many meteorologists agree on the possibility of augmenting both orographic and convective precipitation. Indeed, three decades ago Bergeron (1949) concluded that the main possibility for causing considerable artificial rainfall might be found within certain kinds of orographic cloud systems. This conclusion was based on considerations of the steady and often substantial formation of condensate for an extended period of time in a fixed location and the probable accumulation of 'releasable but unreleased' cloud water at levels with temperatures just below the 0°C isotherm. The basic criterion for determining whether a seeding potential might exist is the natural precipitation efficiency of the clouds – orographic or otherwise. Seeding is clearly not required when the efficiency is high. On the other hand, seeding may or may not be of value when the natural precipitation efficiency is low. The measure of precipitation efficiency is the percentage of condensate which actually reaches the ground. Whilst precise numerical values are difficult to achieve, a useful basis for evaluating precipitation efficiency lies in the comparison of the removal of cloud condensate by vapour growth of ice crystals with the supply of vapour to the cloud. Chappell (1970) has illustrated this idea by comparing for a broad range of cloud temperatures the average rate of formation of condensate, the average rate of consumption of cloud water by ice crystal growth that would occur from natural concentrations of primary ice nuclei and actual average rates of precipitation observed at the ground. His model showed that, with cloud-top temperatures of $-20°C$ or colder, the ice crystals had a rate of water consumption equal to or exceeding the rate of formation of cloud condensate. In the main, such clouds should have a high natural precipitation efficiency with little potential for seeding. The observed actual precipitation corresponded quite closely to the rate at which condensate became available to the clouds.

When cloud-top temperatures were warmer than $-20°C$, the average rate of cloud water consumption from natural concentrations of ice crystals by natural ice nuclei was much lower than the average rate at which condensate became available to the cloud. In these cases considerable quantities of liquid water condensate can be lost to the precipitation process by evaporation which takes place to the lee of the barrier which causes the cloud. For these cases, when natural precipitation efficiency should be low, observed values of ground precipitation were, in fact, much less than the average amount of condensate available. Clearly, a potential for seeding can exist in these cases of 'warm', orographic clouds.

The existence of a potential does not, of course, mean that it is realized. But two sets of observations suggest that realization does occur: first, changes in microphysical characteristics of seeded orographic clouds; and secondly, increased precipitation at the ground. Microphysical changes were recognised in the first field experiments of the 1940s when clouds rapidly glaciated. Changes in ice crystal concentrations, amounts of riming, crystal size and crystal shape have since been documented. In general, concentrations increase, riming decreases, and crystal size decreases. All these changes are consistent with theoretical expectations. Although overall precipitation changes at the ground have been modest, in 'warm' clouds some fairly substantial increases have been observed. For example, at Climax, Colorado precipitation increases of 70–80 per cent have been claimed from clouds with top-temperatures between $-11°C$ and $-20°C$. More frequently, increases of 10–20 per cent are reported.

The second most promising type of precipitation, from a modification point of view, is that from cumulus clouds. In the present context, these clouds are meant to be convective clouds existing in relative isolation, as distinct from convective cells embedded in cyclonic systems or convective cells stimulated by extensive orographic lifting of broad currents of air. The natural precipitation efficiency of these isolated clouds is quite low. Mixing with non-saturated air inhibits the growth of precipitation-sized particles and consequently even the largest clouds – those reaching thunderstorm size – exhibit precipitation efficiencies of only 10 per cent (Braham, 1952). The important question, then, is whether isolated cumuli constitute promising targets for artificial nucleation by virtue of their comparatively low natural precipitation efficiency. Within this context, a major difficulty in assessing possible modification is the enormous natural fluctuations in all variables. Unmodified convective rainfall within seedable situations commonly varies by factors of 10 to 1000, while the largest seeding effects claimed by man have never exceeded a factor of about 3.

In examination of seeding potentiality it has proved useful to subdivide all cumulus clouds into maritime and continental types. Cloud droplet concentrations in these types of cumuli are typically 50 and 400 cm^{-3} respectively. As the cloud liquid-water contents are not greatly different for maritime and continental cumuli, the average droplet radius must be about twice as great in maritime as in continental cumuli. This implies that in maritime cumuli the coalescence process can probably proceed rather rapidly due to a high proportion of initially large droplets; whereas in cumuli containing a continental-type condensation-nucleus population the coalescence process alone would have to operate for a much longer time to result in precipitation. On the basis of this picture, we could conclude that ice nucleants probably offer little potential for stimulating precipitation in maritime cumuli for they can, and evidently do, rapidly develop precipitation by the coalescence process. On the other hand, the same picture suggests that continental cumuli might be artificially modified if one could accelerate particle growth by seeding. This assumes that the continental cumuli have cloud-top temperatures less than 0°C and that natural ice-forming nuclei are so deficient that a substantial part of the cloud water is supercooled.

Attractively simple as the above idea seems to be, it has proved very difficult to test satisfactorily over the last thirty years. Early observational programmes, such as those in both Australia (Warner and Twomey, 1956) and the United States (Braham, Battan and Byers, 1957), were inconclusive. The carefully conducted Project Whitetop, carried out in southern Missouri in 1960–4, revealed a 5–10 per cent increase in radar echo frequency in the region lying just downwind of the seeding area, changing to a decrease of about the same size beyond a downwind distance of around 60–80 km and returning to an increase still farther downwind. This induced rain-shadow effect has become known as the problem of 'robbing Peter to pay Paul', and still requires close scrutiny.

In the tropics a few experiments have claimed increases due to seeding but the general feeling is still that ice nucleants probably are of little effect in initiating precipitation in trade-wind cumuli over the open oceans. In contrast, those cumuli forming over tropical islands may be more promising, but largely because of dynamical rather than micro-physical factors. Howell (1960) has discussed briefly the question of the efficacy of using silver iodide to seed such island cumuli in the tropics, suggesting that buoyancy alteration may be the principal purpose of such seeding.

3.3b Hail suppression

The main reason for wishing to modify hail is its destructiveness. Damage would be reduced of course by preventing hail altogether, but this is not at present feasible. An alternative is to attempt to reduce the size of hailstones which cause most damage. Thus, the present approach to hail modification is to add freezing nuclei which would then compete with natural nuclei for the same finite supply of supercooled water in the cloud. The end-product should be more but smaller hailstones. But this appealing idea has its drawbacks, the main one being as follows. The competition idea implies that the unmodified hail cell is a more or less closed system producing hail at near-capacity rate. In fact, Browning and Foote (1976) have recently demonstrated that large hail cells are both open systems and very inefficient 'hail factories'. They conclude that the addition of nuclei to the main updraught of a storm, which would encourage the production of additional hailstone 'embryos', may increase the number of hailstones of a size equal to that which would have fallen naturally — just the opposite of what is intended. They go on, however, to outline a further way in which seeding may help hail suppression. With the knowledge that a shortage of natural hailstone embryos entering the main storm updraught leads to inefficiency in the production of hail (a natural hail suppression mechanism), Browning and Foote suggest that any action which further decreases the number of embryos should reduce the possibility of large-hailstone growth. One way to achieve this reduction in embryos is to seed the embryo source region on the storm's right flank so as to promote competition during the early growth of the embryos themselves. But first, as Browning and Foote note, we must identify the location of the embryo source more precisely than we are at present able to do. Clearly, the complex and ill-understood interrelations of hailstone and air flow within the hail cloud still provide a significant obstacle to the prospective modifier.

Despite these problems, hail suppression operations have proceeded apace. There are at least eight European, one South American, three African, and eleven Russian projects, quite apart from those in North America. In particular the Russians appear to be convinced that they can mitigate hail and they are doing it routinely, principally by seeding clouds with the aid of rockets. North American visitors to the Soviet operations (Battan, 1977) were impressed by their dedication but remained rather sceptical, particularly in the absence of any breakthroughs in establishing a solid theoretical or physical foundation. In the United States, opinions on the effectiveness of hail suppressions range from 'optimistic' (claims of reductions of 20–48 per cent, but mostly not significant at the 5 per cent level) to 'pessimistic' ('majority of weather modification experts indicate no knowledge of hail suppression capability' (Changnon, 1977)). It is difficult to improve on Changnon's summary: 'Clearly, the current status of hail suppression is in a state of uncertainty. Reviews of the existing results from six recent operational and experimental hail suppression projects are sufficiently suggestive of a hail suppression capability in the range of 20–50 per cent to suggest the need for an extensive investigation by an august body of the hail suppression capability exhibited in these and other programmes' (Changnon, 1977 pp. 26–7).

3.3c Non-meteorological effects of cloud seeding

Despite the many uncertainties involved in cloud seeding, it appears to have increased in scientific respectability in the 1970s. The Russian efforts have been already noted and in the United States a major reappraisal is being made. Although American meteorologists are still cautious in their support for cloud seeding, the number of operations being undertaken is sufficient to stimulate appreciation of possible ecological, social and legal ramifications of seeding.

Over ten years ago a sponsor of a Pennsylvania law designed to curb weather modification said 'Cloud seeding involves silver iodide, and silver iodide, Mr Speaker, is highly poisonous'. He then proceeded to paint a gloomy picture which included 'drastically modified weather conditions . . . total pollution and toxic effects upon all fauna and flora.' His oratory succeeded and the bill became law, yet recent analyses (Klein and Molise, 1975) have shown that the silver concentrations in grasslands where seeding has taken place show no significant changes. In parallel with such ecological studies, Davis (1975) has shown that many states in USA have now drafted laws which allow close governmental control over the environmental consequences not only of cloud seeding but of all of weather modification. The social aspects of weather modification are also now being considered (Hass, 1973; Lansford, 1973), the general tone of the discussion being that even if cloud seeding becomes a reliable technology, the public rather than scientists or government will decide whether they wish it to be used.

3.4 UNCONSCIOUS MODIFICATION OF PRECIPITATION

Whereas cloud-seeding occupies a well-known, if not widely accepted position within meteorology, man's inadvertent effects upon precipitation are largely unappreciated. But the last two decades have seen a fairly steady increase in evidence of such effects and this work is reviewed in the remainder of this chapter. Initially however, it is worthwhile reiterating the requirements for precipitation because certain factors figure prominently in the assessment of inadvertent effects. Precipitation results from the cooling (usually by uplift) of moist air so that condensation, coalescence, and the ice-crystal mechanisms occur. Clearly, any increases in water vapour content, uplift and condensation, and ice nuclei would tend, at least in a qualitative sense, to favour precipitation formation. Because the amounts of water vapour, nuclei, and, to a lesser extent, the degree of uplift depend on the nature of the earth's surface, man-made changes in this surface may affect these three factors and thus inadvertently affect precipitation. Thus man's unconscious effects are indirect, in complete contrast to his modifications by cloud seeding.

Two major ways in which man modifies the landscape are by planting or clearing of trees and by building urban areas. In the nineteenth century, large scale deforestation was popularly thought to decrease precipitation. Inversion of this idea leads to schemes for increasing regional precipitation by planting trees, and at least one argument has seriously suggested that the states of Oregon and Washington attest to the rain-producing influence of extensive forests. At present however, little or no evidence exists for significant changes in precipitation amounts due to the presence or absence of a forest.

The situation for urban areas is rather different. As early as the 1930s, suggestions emanated from central Europe that urban areas may occasionally induce an increase in precipitation. The first analyses were essentially climatological, relating in a spatial sense mean annual or seasonal values of rainfall amounts or frequency to the form of the urban area (e.g. Changnon, 1961; Huff and Changnon, 1973). In many of these studies, increases in annual amounts of about 10 per cent were indicated and three possible explanatory factors were suggested: increased numbers of condensation and ice nuclei; increased absolute humidities in cities; and increased uplift over cities due to both thermal and mechanical effects. As a result of detailed case studies and the massive observational Metropolitan Meteorological Experiment (METROMEX) (Changnon, Huff and Semonin, 1971) we are now more certain of the absolute and relative effects of the three factors noted above. Before considering each in turn it is worth noting that METROMEX also provided abundant data which essentially sharpened the problem. A detailed analysis of 5-minute rainfall amounts over St. Louis (Huff and Schickedanz, 1974) showed that in 'rain cells' which had passed over the urban-industrial region of the town, the average rainfall volume was 216 per cent greater than in cells outside the urban area. It is important to note that this large increase applied only to distinct rain cells in two summers and it should not be thought of as a representative result for all rainfalls in all seasons over a long period. Nevertheless, the increase leaves little doubt as to the reality of an urban effect – albeit an effect which operates spasmodically.

The problem of urban-induced precipitation becomes more tractable if attention is restricted to convective precipitation. As argued previously (Atkinson, 1968, 1969) convective cloud dynamics often have a clear link with thermal sources at the ground and as such would be expected to respond to changes in surface cover. Within this context we can now profitably review the three factors which affect urban precipitation.

The effects of increases in the concentrations of condensation and ice nuclei are not easy to determine. Apart from being difficult to measure, the importance of such nuclei in determining precipitation amounts is probably far less than that of the air motions within clouds – particularly the updraughts (Battan, 1965). Recent evidence has shown that condensation nuclei numbers increased substantially in the towns of Buffalo, St Louis, and Seattle, but interpretations differ on the effects that these increases had on the spectrum of sizes of cloud droplets. In Buffalo and St Louis respectively, Kockmond and Mack (1972) and Fitzgerald and Spyers-Duran (1973) found that the droplet spectrum narrowed markedly in the industrial plume. This means that there were more cloud droplets, but that they were smaller than required for growth of precipitation-sized particles. These findings support the laboratory results of Gunn and Phillips (1957). In contrast, Hobbs, Radke and Shumway (1970), working in Seattle, found that the droplet spectrum broadened due to the initiation of many droplets of diameter greater than $30\mu m$. Such sizes are ideal for the operation of a very efficient coalescence mechanism – thus rapidly producing precipitation-size drops. Braham (1974) was of the opinion that the instrument used in METROMEX was inadequate to measure droplets of $30\mu m$ diameter, but felt that they must have been present as radar observations showed preferential echo growth over the St Louis area. Braham argued that for this to happen, coalescence must have been a powerful mechanism and this, in turn, requires the large droplets of the kind measured by Hobbs and his colleagues. Later measurements from METROMEX (Braham and Spyers-Duran, 1974) have revealed lower ice nuclei concentrations over St Louis than in the air upwind of the city. Clearly we are not yet in a position to reach firm conclusions about the effects of urban areas on the concentrations of ice and condensation nuclei and, in turn, their effects on precipitation amounts.

Measurements of humidities in urban areas are as scarce as those of condensation and ice nuclei but a notable contribution to our knowledge of the former is provided by

Chandler's (1967) observations of relative and absolute humidities in Leicester, England. They are particularly attractive in the present context because Chandler found that water vapour pressures on three August 1966 nights were 2 mb higher within the urban area than in the surrounding rural areas. The measurements by Bornstein, Lorenzen and Johnson (1972) in New York confirmed Chandler's results and showed that urban excesses of absolute humidities may extend up to a height of 700 m. Such results, if they have general applicability, imply that urban air has a lower lifted condensation level and a lower level of free convection and thus is more permissive of cloud and precipitation growth than rural air.

Important as nuclei and humidities are, the prime factor influencing precipitation amounts is the nature of the cloud updraught. In general increases in updraught area, depth, and speed lead to increases in precipitation at the ground. Urban areas may affect these updraught characteristics by virtue of their thermal and mechanical influence on airflow. Each is considered below.

The thermal effect of the city upon both stationary and moving thunderstorms has been analysed by Atkinson (1970, 1971). Only the stationary case is used here for exemplification. On 21 August 1959 localized, heavy precipitation fell over London, England (Fig. 3.1), with a maximum amount of 68 mm. Radar evidence clearly showed the growth of deep clouds over the urban area, and this was strongly suggestive of an urban effect. These storm clouds were preceded by surface temperatures of 27–29°C in the city, the highest temperatures anywhere in south-east England (Fig. 3.2). The lapse-rate was superadiabatic in the lowest 500 m of the atmosphere and perhaps, in the Finsbury, Shoreditch, Bethnal Green area where surface temperatures reached 29°C, auto-convective instability existed in the lowest 100 m. Such strong instability encourages vigorous

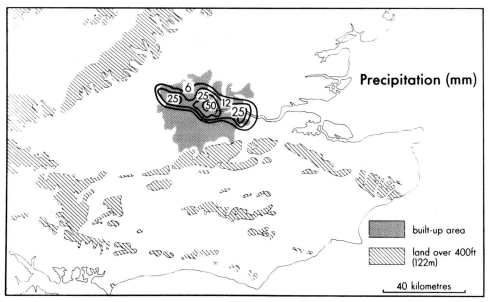

FIG. 3.1 TOTAL PRECIPITATION (mm) OVER SOUTH-EAST ENGLAND, 21 AUGUST 1959

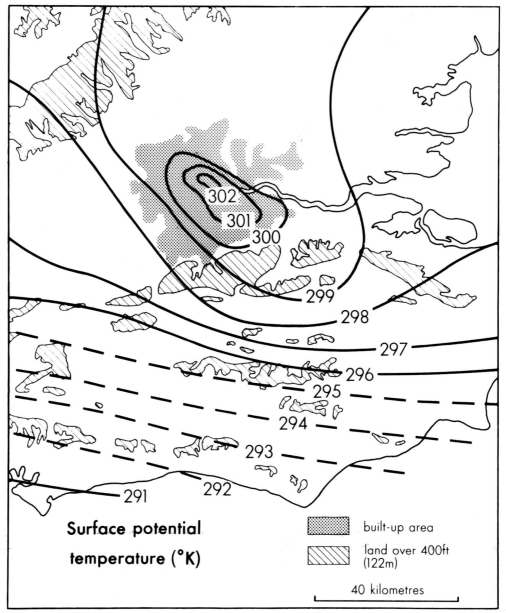

FIG. 3.2 SURFACE POTENTIAL TEMPERATURE (K) 1200 GMT, 21 AUGUST 1959
Potential temperature is the temperature taken on by an air parcel brought adiabatically to a pressure of 1000 mb

vertical motion and near surface values of the wet-bulb potential temperature of 20°C (Fig. 3.3) meant that air rising adiabatically would reach equilibrium temperature at a height of 11 km. This was in good agreement with the radar observation of echo tops at

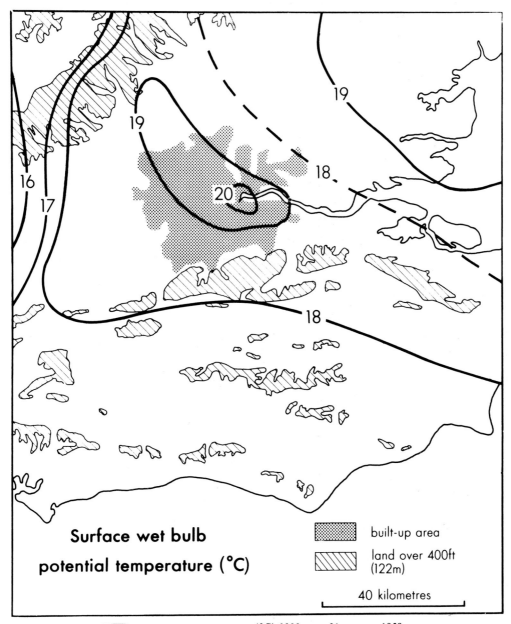

FIG. 3.3 SURFACE WET-BULB POTENTIAL TEMPERATURE (°C) 0900 GMT, 21 AUGUST 1959

about 10·6 km in the early afternoon. There appears to be little doubt that in the initially calm atmosphere of that August afternoon, London's heat island played a significant role in releasing the instability which led to such heavy precipitation.

The mechanical effect of urban areas upon airflow manifests itself in two ways: first, as an obstacle to flow; and secondly, by causing frictional convergence. Angell, Hoecker, Dickson and Pack (1973) found that in strong winds of about 13 m s^{-1} Oklahoma City acted as a barrier to flow and induced upward velocities to air parcels of up to 70 cm s^{-1} at the 400 m level over parts of the city. The upward velocities varied in magnitude with time of day, being strongest between 1200 and 1800, weakest between 1800 and 2100 and between those extremes from 0900 to 1200 local time. Observations of flow around the city also suggested a barrier effect which was not fully compensated by vertical motion. Angell *et al.* also suggested that frictional convergence occurred, a theme more fully developed by Ackerman (1974).

Ackerman found that in winds of less than 5 m s^{-1} pronounced perturbations occurred in the horizontal flow over St Louis on two August afternoons. In the lowest 500–600 m of the atmosphere, airflow took on an anticyclonic curvature as it approached the city centre and then cyclonic curvature as it moved toward the downwind city edge. The result of these perturbations was a convergence of air over the city and this necessitated uplift. Ackerman calculated upward velocities of 2–6 cm s^{-1}, apparently small, but, as she points

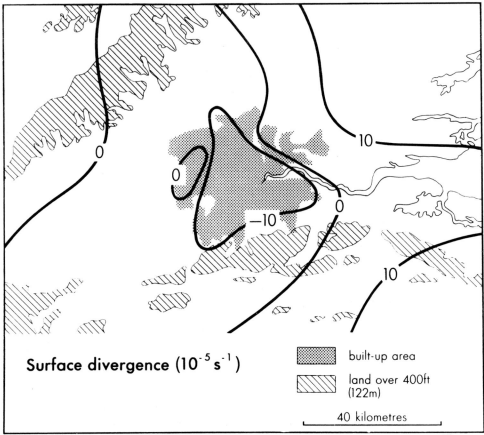

FIG. 3.4 DIVERGENCE OF SURFACE WIND FIELD (UNITS OF 10^{-5}s^{-1}) IN SOUTH-EAST ENGLAND 0900 GMT, 1 SEPTEMBER 1960

out, 'the mean ascent rate could represent the net effect of a number of thermal bubbles or plumes of restricted areal extent but more intense vertical motion, rather than sustained flow. Indeed, vertical velocities deduced from the ascent rates of pibals indicate that local updraughts of 0·5 to 1 m s⁻¹ did occur periodically at many sites . . .' (p. 234).

Applying these principles to a London case study, Atkinson (1975) showed that the precipitation of 1 September 1960 was partially due to frictional convergence over the city. Figure 3.4 shows an area of strong convergence (10^{-4} s⁻¹) lying over the bulk of the urban area with extreme values of 2×10^{-4} s⁻¹ in the Greenwich area. Such values of convergence, if they existed over a substantial depth of the lowest kilometre as indicated by Ackerman, would result in an uplift of air of about 10 cm s⁻¹. Consequently, to lift the surface layers through a distance of 1 km would take 10^{4} s or about 2½ hours. By 1200 GMT on 1 September the convergence had been going on long enough to lift the bottom 500–1000 m of air to saturation and to a state of absolute instability. It was that air which was drawn into the convective clouds which moved over the city and gave heavy precipitation (Fig. 3.5).

3.5 CONCLUSION

Man's effects upon precipitation have been suspected for only fifty years, and verification of these suspicions remains incomplete. Yet, as this chapter has attempted to show, such evidence as is increasingly available suggests that both conscious and unconscious effects are real. The magnitude and frequency of occurrence of the effects are small but,

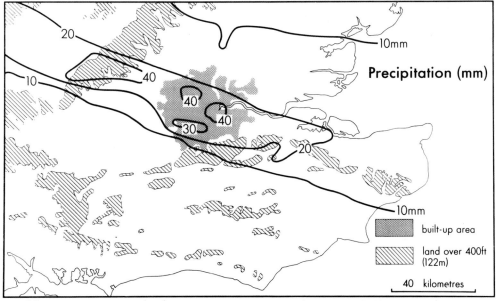

FIG. 3.5 TOTAL PRECIPITATION (mm) OVER SOUTH-EAST ENGLAND, 1 SEPTEMBER 1960

nevertheless, may have a significant impact on man's other activities. Thus an artificially induced 10 per cent increase in precipitation in the right place at the right time could be of great value to agriculturalists in areas of low rainfall, such as Australia, Israel and the American Mid-West. Conversely a very heavy storm, perhaps triggered by an urban area (Atkinson, 1977), may result in flooding, extensive damage to property and even loss of life. In both cases, the precipitation amounts due to man's activities are small compared to the 'natural' precipitation over a period of one year, but their immediate impact is of prime importance. The social, economic, and legal implications of both forms of precipitation modification will no doubt ensure a healthy future for scientists interested in the mechanisms of precipitation formation.

REFERENCES

ACKERMAN, B,. 1974, 'Wind fields over the St Louis metropolitan area.' *Jnl. Air Poll. Cont. Assoc.*, 24, pp. 232–6.

ANGELL, J. K., HOECKER, W. H., DICKSON, C. R. and PACK, D. H., 1973, 'Urban influences on a strong day time air flow as determined by tetroon flights.' *Jnl. Appl. Met.*, 12, pp. 924–36.

ATKINSON, B. W., 1968, 'A preliminary examination of the possible effect of London's urban area on the distribution of thunder rainfall, 1951–60.' *Trans. Inst. Brit. Geog.*, 44, pp. 97–118.

1969, 'A further examination of the urban maximum of thunder rainfall in London, 1951–60.' *Trans. Inst. Brit. Geog.*, 48, pp. 97–119.

1970, 'The reality of the urban effect on precipitation — a case study approach', in 'Urban Climates', *Tech. Note No. 108*, World Meteorological Organization, pp. 342–60.

1971, 'The effect of an urban area on the precipitation from a moving thunderstorm.' *Jnl. Appl. Met.*, 10, pp. 47–55.

1975, 'The mechanical effect of an urban area on convective precipitation.' *Occas. Pap. No. 3*, Dept. Geog., Queen Mary College, University of London.

1977, 'Urban effects on precipitation: an investigation of London's influence on the severe storm in August 1975.' *Occas. Pap. No. 8*, Dept. Geog., Queen Mary College, University of London.

BATTAN, L. J., 1965, 'Some factors governing precipitation and lightning from convective clouds.' *Jnl. Atmos. Sci.*, 22, pp. 79–84.

1977, 'Weather modification in the Soviet Union – 1976.' *Bull. Amer. Met. Soc.*, 58, pp. 4–19.

BERGERON, T., 1949, 'The problem of artificial control of rainfall on the globe. I. General effects of ice-nuclei in clouds.' *Tellus*, 1, pp. 32–50.

BORNSTEIN, R. D., LORENZEN, A., and JOHNSON, D., 1972, 'Recent observations of urban effects on winds and temperatures in and around New York City.' *Preprints Conf. Urban Environment and Second Conf. on Biometeorology, Amer. Met. Soc.* (Boston), pp. 28–33.

BRAHAM, R. R., 1952, 'Water and energy budgets of the thunderstorm and their relation to thunderstorm development.' *Jnl. Met.*, 9, pp. 227–42.

1974, 'Cloud physics of urban weather modification — a preliminary report.' *Bull. Amer. Met. Soc.*, 55, pp. 100–6.

BRAHAM, R. R., BATTAN, L. J., and BYERS, H. R., 1957, 'Artificial nucleation of cumulus clouds.' *Meteorological Monog., Amer. Met. Soc.*, 11, pp. 47–85.

BRAHAM, R. R., and SPYERS-DURAN, P., 1974, 'Ice nucleus measurements in an urban atmosphere.' *Jnl. Appl. Met.*, 13, pp. 940–5.

BROWNING, K. A., and FOOTE, G. B., 1976, 'Airflow and hail growth in supercell storms and some implications for hail suppression.' *Quart. Jnl. Roy. Met. Soc.*, 102, pp. 499–534.

CHANDLER, T. J., 1967, 'Absolute and relative humidities in towns.' *Bull. Amer. Met. Soc.*, 48, pp. 394–9.

1970, *The Management of Climatic Resources.* Inaugural Lecture, University College London (H. K. Lewis and Co., London).

CHANGNON, S. A., 1961, 'A climatological evaluation of precipitation patterns over an urban area.' In *Air Over Cities.*, Sec. Tech. Rep. A62–5., Lab. of Engin. and Phys. Sci. of Div. of Air Poll., US Dept of Health, Educ., and Welfare (Public Health Service, Washington, D.C.), pp. 37–67.

1977, 'On the status of hail suppression.' *Bull. Amer. Met. Soc.*, 58, pp. 20–8.

CHANGNON, S. A., HUFF, F. A., and SEMONIN, R. G., 1971, 'METROMEX: an investigation of inadvertent weather modification.' *Bull. Amer. Met. Soc.*, 52, pp. 958–68.

CHAPPELL, C. F., 1970, 'Modification of cold orographic clouds.' *Atmos. Sci. Pap. No. 173*, Colorado State Univ.

DAVIS, R. J., 1975, 'Legal response to environmental concerns about weather modification.' *Jnl. Appl. Met.*, 14, pp. 681–5.

FITZGERALD, J. W., and SPYERS-DURAN, P. A., 1973, 'Changes in cloud nucleus concentration and cloud droplet size distribution associated with pollution from St Louis.' *Jnl. Appl. Met.*, 12, pp. 511–16.

GUNN, R., and PHILLIPS, B. B., 1957, 'An experimental investigation of the effect of air pollution on the initiation of rain.' *Jnl. Met.*, 14, pp. 272–80.

HASS, J. E., 1973, 'Social aspects of weather modification.' *Bull. Amer. Met. Soc.*, 54, pp. 647–57.

HOBBS, P. V., RADKE, L. F., and SHUMWAY, S. E., 1970, 'Cloud condensation nuclei from industrial sources and their apparent influence on precipitation in Washington State.' *Jnl. Atmos. Sci.*, 27, pp. 81–9.

HOWELL, W. E., 1960, 'Cloud seeding in the American tropics.' *Physics of Precipitation, Geophys. Monog. No. 5.*, Amer. Geophys. Union, pp. 412–23.

HUFF, F. A. and CHANGNON, S. A., 1973, 'Precipitation modification by major urban areas.' *Bull. Amer. Met. Soc.*, 54, pp. 1220–33.

HUFF, F. A. and SCHICKEDANZ, P. T., 1974, 'METROMEX: rainfall analyses.' *Bull. Amer. Met. Soc.*, 55, pp. 90–2.

KESSLER, E., 1973, 'On the artificial increase of precipitation.' *Weather*, 28, pp. 188–94.

KLEIN, D. A. and MOLISE, E. M., 1975, 'Ecological ramifications of silver iodide nucleating agent accumulation in a semi-arid grassland environment.' *Jnl. Appl. Met.*, 14, pp. 673–80.

KOKMOND, W. C. and MACK, E. J., 1972, 'The vertical distribution of cloud and Aitken nuclei downwind of urban pollution sources.' *Jnl. Appl. Met.*, 11, pp. 141–8.

LANSFORD, H., 1973, 'Weather modification: the public will decide.' *Bull. Amer. Met. Soc.*, 54, pp. 658–60.

MASON, B. J., 1971, *The Physics of Clouds,* (OUP, Oxford, 2nd edn.).

SCHAEFER, V. J., 1946, 'The production of ice crystals in a cloud of supercooled water droplets.' *Science*, 194, pp. 457–9.

VONNEGUT, B., 1949, 'Nucleation of supercooled water clouds by silver iodide smokes.' *Chem. Rev.*, 44, pp. 277–89.

1950, 'Experiments with silver iodide smokes in the natural atmosphere.' *Bull. Amer. Met. Soc.*, 31, pp. 151–7.

WARNER, J. and TWOMEY, S., 1956, 'The use of silver iodide for seeding individual clouds.' *Tellus*, 8, pp. 453–9.

4

Man-Modified Climates

WAYNE R. ROUSE
McMaster University

4.1 INTRODUCTION

The evidence for substantial climatic modification due to man's activities has increased dramatically over the last decade and there is little doubt that the impact will accelerate in the near future. Climatic modification is first felt, or at least first recognized, in the continental areas, where human enterprise is concentrated, but the effects ultimately include the world oceanic areas because of the global nature of the earth-atmospheric energy exchange system and the hydrological cycle. In this chapter we can only develop the basic nature of man's impact on climate and will not attempt to be comprehensive. Climatic modification is a subject about which little is known, but much is suspected. In addition, the available evidence is often conflicting. On a global scale, the ability to model man's impact on climate is still in a primitive scientific state. However, notable efforts are being made in the modelling field. A summary of the 'state-of-the art' of climatic modelling is provided by Schneider and Dickinson (1974).

This discussion aims to introduce the principles of the radiation balance at the earth's surface, the surface energy or heat balance, patterns of evaporation, and the behaviour of soil water, for it is with these basic climatic systems that man's changes in the landscape interact. The potential large-scale impacts involve the following: water diversion schemes for hydro-electric generation and irrigation; forest removal, both deliberate and inadvertent; intensified agriculture and livestock herding in climatically marginal lands; increased urbanization and the development of urban agglomerates; and atmospheric changes in carbon dioxide and dust content.

4.2 BASIC PRINCIPLES

The radiation balance (see also Chapter 2) is a statement of the conservation of radiant energy at the earth's surface and it is expressed in terms of the solar radiation received at the surface $K\!\downarrow$, the solar radiation reflected $K\!\uparrow$, the long wave infrared radiation received from the earth's atmosphere $L\!\downarrow$ and the long wave radiation emitted from the earth's surface $L\!\uparrow$, in the form

$$Q^* = K\!\downarrow - K\!\uparrow + L\!\downarrow - L\!\uparrow \qquad (4 \cdot 1)$$

where Q^* is termed the net radiation. An alternative form of the radiation balance is expressed as

$$Q^* = K\!\downarrow(1-\alpha) + L\!\downarrow - L\!\uparrow \qquad (4 \cdot 2)$$

where $\alpha = K\!\uparrow/K\!\downarrow$ and is known as the surface albedo. The magnitude of $K\!\downarrow$ varies with latitude, time of year and time of day and with atmospheric transparency. The latter is most strongly influenced by cloud cover and atmospheric dust content. The surface albedo varies from very low values, usually less than 0·06 over water and burned surfaces, to very high values of up to 0·90 over fresh snow surfaces. Most vegetated surfaces have an albedo between 0·12 and 0·25, whereas unvegetated surfaces such as deserts have albedos rising as high as 0·45. Albedo represents one of the major elements in the radiation balance that can be influenced by terrain modification. The magnitude of long wave radiation varies with the surface temperature and efficiency of the radiator. The opaque earth's surface is a very efficient radiator, whereas the gaseous atmosphere is much less efficient. As a result the long wave balance $L\!\downarrow - L\!\uparrow$ is almost always negative. If the surface is hot and the atmosphere clear and cool, a long wave balance of large negative magnitude is achieved. If the atmosphere is warmer than the surface, or is cloudy, the negative balance is small. In addition to clouds, atmospheric water vapour and carbon dioxide are effective in absorbing $L\!\uparrow$. This in turn raises the temperature of the atmosphere and enhances the magnitude of $L\!\downarrow$. This phenomenon, by analogy, is known as the 'greenhouse effect' and is another means by which climate may be modified as will be dicussed later.

The net radiation Q^* is positive during daylight hours, when the absorbed solar radiation is larger than the net long wave loss, and negative at night when there is only net long wave loss. In similar fashion, beyond the Arctic Circle Q^* can be positive for the full diurnal period at the summer solstice and conversely negative for the full day at the winter solstice. On a yearly basis Q^* is positive at most latitudes except glaciated high latitude and high altitude zones.

The net radiant energy, whether positive or negative, is balanced by flows of heat energy between earth and atmosphere. The major forms of this heat flow, also known as heat flux, are the latent heat flux LE which occurs during evapotranspiration, where L is the latent heat of vaporization of water and E is the amount of water evapotranspired; the sensible heat flux H, which is the vertical movement of warmed air between the earth's surface and atmosphere; and the subsurface heat flux G, which is the vertical movement of heat between the surface and deeper layers in soil or water. The balance equation

WAYNE R. ROUSE

known as the energy or heat balance is expressed as

$$Q^*=LE+H+G \qquad (4\cdot3)$$

Figure 4.1 shows typical forms of the energy balance. The energy balance of a moist
snow-free terrestrial surface is shown in Figure 4.1A with positive net radiation of large
magnitude denoted by the long downward-directed arrow. All other energy flows are
directed away from the surface, evapotranspiration being of largest magnitude. Figure
4.1B shows that at night the net radiation is negative and small, and that all other energy
movements are toward the surface, with the sensible heat flux normally being the largest.
The water vapour in the downward directed latent heat flux condenses at or near the
cooling surface in the form of dew or radiation fog. Over a totally arid desert shown in
Figure 4.1C almost all of the net radiation is balanced by the sensible heat flux since there
is no water to evaporate. Over a desert oasis, where soil water is available for evaporation,
the situation is different as seen in Figure 4.1D. The wind blows hot dry air from the
surrounding desert into the oasis. The temperature of this air is much higher than the cool

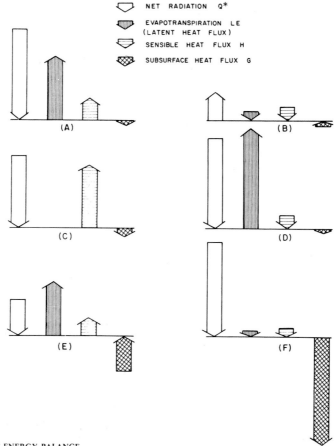

FIG. 4.1 TYPES OF ENERGY BALANCE
(A) moist surface — daytime (B) nightime (C) desert surface — daytime (D) desert oasis — daytime (E) warm
ocean current in winter (F) cold ocean current in summer

oasis surface so the sensible heat flux is directed toward the surface. With this extra heat energy the latent heat flux is enhanced. Thus larger amounts of water evapotranspire than would be possible from the input of radiational energy alone. This type of energy balance pattern is so characteristic that it has earned the name 'oasis effect'. It applies, however, to more than a desert oasis and will occur whenever a moist surface lies downwind of a warmer drier environment. It is especially important where artificial reservoirs are created in arid and semi-arid regions. Figures 4.1E and 4.1F are designated as oceanic situations, but also apply to lakes in the middle and high latitudes. A good example of the warm ocean current in winter (Fig. 4.1E) is water in the Gulf Stream. Although the net radiation is positive it is small in magnitude. The latent and sensible heat fluxes are relatively large, however, due to oceanic heat which originated at subtropical latitudes in the Sargasso Sea and Gulf of Mexico. The importance of this winter warming effect to western Europe is well known. It is an effect which also applies, in the middle and high latitudes, in the fall season, when the large amount of heat energy stored in lakes is released to the atmosphere. This is especially noteworthy in North America, where it enhances heavy winter snowfalls to the lee of the Great Lakes. The cold-current regime in summer (Fig. 4.1F) operates in reverse, creating downward-directed latent and sensible heat fluxes with all energy inputs being used to warm the water. The effects can be large-scale and include the perpetuation of fog over the Labrador current, especially notable around Newfoundland, and the intensification of the desert climates of north Africa, southwest Africa and Peru-northern Chile due respectively to the Canaries, Benguela and Humboldt cold currents off the coast.

Over terrestrial surfaces the magnitude of the latent heat flux is a function of the availability of soil water, the available radiant heat energy, the temperature and dryness of the atmosphere and the rigour of turbulent mixing in the lower atmosphere.

The amount of water in the soil is most usefully expressed as a percentage volume. Thus if the volumetric moisture content is 50 per cent it means, for example, that a cubic metre of soil would contain half a cubic metre of water. This would represent a very wet or saturated soil, probably in a swamp environment. For unsaturated soils it is convenient to treat the available soil water as ranging between two reference points known as field capacity and permanent wilting point. These two points can be envisaged as follows. When a plant-covered soil is saturated by rainfall or irrigation, there is initially a strong downward movement of this water toward the water table. After a day or so this flow largely ceases because the suction force exerted by soil particles balances the effects of gravitational attraction. The amount of water remaining in the soil at this near-equilibrium is known as the field capacity and represents water which is available for evapotranspiration. A subsequent period of evapotranspiration will remove this water and, if sustained long enough, the soil will become very dry. Eventually a state is reached where the suction forces exerted on the soil water by the soil particles are equal to the maximum suction that a plant can exert through its roots and no more water can be withdrawn. The plant dies and this represents the permanent wilting point. The volume of available water between field capacity and permanent wilting point is highly variable. Fine-grained soils such as clays and clay loams have a large interface area and exert strong suction forces. They hold abundant water at field capacity and are still fairly moist when the permanent wilting point is reached. For these soils a typical range of soil moisture content between the two reference points is 45 to 15 per cent. This gives a soil moisture content of 30 per cent available for evapotranspiration and the life support of plants. In contrast, a soil comprised primarily of coarse-grained sands, with small interface area, will have a moisture range of between 20 and 5 per cent. This gives an available soil moisture content of only 15 per cent. The greater the organic content of coarse-grained soils, the more water that is available to sustain plant growth and also the more water than can enter the atmosphere with the potential for increasing local precipitation. Poor land-use practices through overgrazing, overcropping or wind erosion can change the soil water retention capabilities and have a multiple effect on climate. Examples where this appears to be the case will be considered later in the chapter.

This discussion of basic principles may now be used as a basis for discussing the climatic modifications associated with man-made lakes, forest removal, desertification, and urbanization and industrialization.

4.3 MAN-MADE LAKES

In a hungry search for hydro-electric power and irrigation water many of the world's rivers have been dammed to create artificial lakes, often of considerable size, and it is a fair guess that by early in the twenty-first century, all of the world's rivers with any potential for power or irrigation will have been dammed.

It has been estimated by Fels and Keller (1973) that 315 large man-made reservoirs had been created by 1973 and that this involved a surface area in excess of 200 000 km². The geographical distribution of these is shown in Table 4.1. It is evident that impoundments are found in all geographical areas and in a variety of climatic types, including tropical rainy regimes, subtropical deserts and subarctic zones. The USSR has the largest total area of reservoirs and also the largest average reservoir size. Since the

Table 4.1

LARGE MAN-MADE LAKES

Location	Number	Approx. surface area (km²)	Av. surface area/reservoir (km²)	Av. depth (m)	Percent of world total
USSR	14	79477	5677	14	37
Canada	51	40839	801	15	19
USA	112	30175	269	13	14
Africa	17	27740	1623	23	13
Latin America	31	14797	477	20	7
Asia except USSR	18	10030	557	24	5
India	25	6522	261	18	3
Europe	9	2301	256	23	1
Australia	12	2284	190	25	1
World	289	214165	741	20	100

Source: Fels and Keller (1973)

climatic impact of man-made lakes differs with the climatic zone, it is convenient to consider the tropical-subtropical regions and temperate-high latitude zones separately.

In flooding a land area in the low latitudes, one of the immediate effects is to increase the net radiation. This arises for two reasons which are implicit in equation (4.2). First, the albedo is decreased to around 6 per cent. This may be from a previous value of 24 per cent for an unvegetated desert surface, 16 per cent for a savanna grassland surface or 21 per cent for a tropical forest. Secondly, $L\uparrow$, the long-wave radiation loss from the surface, is substantially less because of much cooler daytime temperatures in the reservoir. These two effects can increase $Q*$ by up to 25 per cent when a man-made lake replaces a high albedo arid surface. This provides more net energy for evaporation from the water surface and on a local scale more water is evaporated into the atmosphere. In wet tropical areas the influence will be minimal. In arid zones, however, the effect is substantial. The larger net radiation combined with the oasis effect (Fig 4.1D) will strongly enhance the reservoir evaporation. A classic study of the evaporation from an arid zone impoundment is the case of Lake Mead as reported by Harbeck et al. (1958). Lake Mead was created in 1935 by damming the Colorado River near Boulder, Colorado. With a surface area of 660 km^2 and average depth of 180 m it is the largest man-made lake in the United States. The climate in the area is semi-arid with average annual precipitation less than 130 mm and an annual mean temperature of 19°C. In the study by Harbeck et al. (1958) it was found that the evaporation used 16 per cent more energy than was provided by net radiation and about 30 per cent more energy than would be expected in a more humid environment. A similar oasis effect occurs in Lake Nasser, Egypt, which has a surface area of 5237 km^2 and where the precipitation is almost nil (Dekker, 1972), and in other man-made lakes in subtropical arid and semi-arid zones. Where dry land irrigation accompanies the impoundment the effects are accentuated. Examples of the effects of dry land irrigation on the energy balance are provided by Flohn (1971) in a study of oases in the Tunisian semi-desert zone. Although the irrigation water comes from subterranean sources, rather than river impoundments, the effects will be similar. Figure 4.2 shows that the strength of the oasis effect varies with oasis type but that in all cases the effect is a higher net radiation and enhanced evaporation in comparison to the surrounding dry lands. In the case of the humid oasis the use of advected heat energy (negative H) for evapotranspiration greatly exceeds the heat input from net radiation. There is no subsurface heat flow G in Figure 4.2, since for daily or longer periods it tends toward zero. Lvovich (1969) estimated that evaporated irrigation water represents a decrease of about 5 per cent in the run-off from land areas and an addition of 2 per cent to the total annual evaporation from terrestrial surfaces. He also estimated that by the beginning of the twenty-first century the demand for irrigation water will have doubled and that, in combination with increased industrial demands, the amount of evaporating water due to water diversion will be three times as great as in 1965. This change in the world's water balance could well have global climatic effects, a possibility which is discussed in the SMIC (1971) report.

In the middle and high latitudes the creation of artificial lakes imposes climatic effects which are tied to the seasons. In summer time there is an increase in net radiation of similar magnitude to low latitude reservoirs. In winter, however, the frozen snow-covered reservoir has a high albedo and correspondingly small net radiation. This effect becomes notable in spring when land areas are free of snow and are assuming a summertime

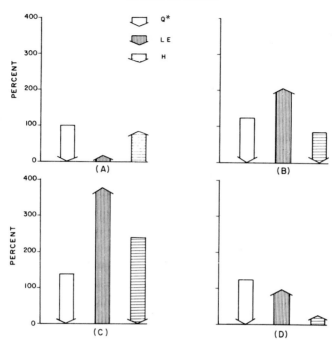

FIG. 4.2 ENERGY BUDGETS OF TUNISIAN OASES COMPARED TO THE SURROUNDING SEMI-DESERT
Magnitudes are in terms of $Q^* = 100$ per cent for the semi-desert. (A) semi-desert (B) typical oasis (C) humid
oasis (D) dry oasis. (Source: Flohn, 1971. Data used with permission of the World Meteorological Organization)

radiation balance, whereas the reservoirs maintain a wintertime regime. In the autumn,
however, the situation is reversed with a substantial time lag in freezing of the reservoirs,
so that open water bodies are surrounded by snow-covered terrestrial surfaces. It is an
irony of geography that, in the subarctic and arctic regions, when the solar energy is large
as the sun approaches its zenith at the summer solstice, the ground remains snow-covered
thus maintaining a high albedo and the absorbed solar energy is small. This effect is
accentuated by the time lag in the thawing of natural and artificial lakes. The process is
not fully reversible since the longer open-water state in the autumn occurs during a period
of small solar radiation. The evaporation cycle for lakes is similarly displaced temporally.
An example of this is Lake Diefenbaker in Canada, a reservoir having a surface area of
540 km² and an average depth of 24 m, created by the damming of the South
Saskatchewan River to the west of Regina. The water temperature lags behind the air
temperature during both the spring warming cycle and fall cooling cycle, and reaches its
maximum temperature in late August to early September. The evaporation rate also peaks
at the same time and there is still substantial evaporation through to the end of November
(Buckler, 1973). Figure 4.3 plots the energy balance for Lake Diefenbaker during the ice
free period using data from Cork (1974) and the patterns can be accepted as typical of
lakes in the temperate and subarctic regions. Whereas the net radiation peaks in the
months of June and July, evaporation peaks in August, September and October. The
sensible heat flux is directed downwards in May and June from the warm atmosphere to

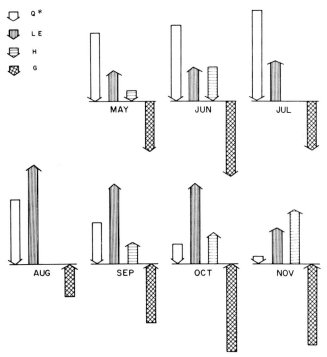

FIG. 4.3 THE ENERGY BALANCE OF LAKE DIEFENBAKER IN 1973
LE and H for June are estimated. Magnitudes are in terms of $Q^* = 100$ per cent in July. (Prepared from data by Cork, 1974)

the cooler lake waters with a reversal of the situation during the months from September to November. There is no sensible heat flux in July and August because water and air temperatures are the same. The heat storage in the lake similarly shows a seasonal flip flop and in the fall the heat released and used in the latent and sensible heat fluxes greatly exceeds the magnitude of the net radiation.

High evaporation certainly increases rainfall and snowfall downwind from natural lakes in the autumn and early winter. This is a well documented phenomenon in the Great Lakes of North America where heavy snowfall results to the lee of the lakes, as documented by Brown et al. (1968) and McVehil and Peace (1965), and in Lake Baikal, USSR, as reported by Kornienko (1969) and discussed by Nemec (1973). Evidence for climatic effects from a large artificial reservoir in the middle latitudes comes from the Rybinsk reservoir built on the upper Volga in 1941 a distance of 320 km north of Moscow. The reservoir is very large and shallow with a surface area of 4550 km[2] and an average depth of 5·6 m. Butorin et al. (1973) report the following climatic effects. In spring, the shoreline areas with onshore winds are colder than average, whereas in the autumn they are warmer than average, with an average monthly temperature difference of from 2 to 5°C compared to surrounding areas. The same areas have from 15 to 30 per cent less precipitation in the spring and between 90 and 120 per cent more in the autumn. Butorin *et al.* report that the maximum distance for this climatic effect extends to 15 km from the shoreline.

One of the most ambitious diversionary schemes for hydro-electric power development is taking place in northern Quebec in Canada and it has elicited a heated emotional reaction. The project ultimately will harness all of the major rivers flowing into the east side of James and Hudson Bays and flowing north into Ungava Bay and will affect an area of about 435 000 km^2. The five largest reservoirs which are being created in the initial stages of the project will comprise a surface area of about 8500 km^2 which amounts to a third of the size of Lake Ontario and a quarter of the size of Lake Erie. It is only possible at this stage to speculate on the climatic effects, but there are potential problems that should be of concern in a geographical area where there is little scientific knowledge. First is that local downwind areas will experience an increased early winter snowfall, although this is a period when most of the snow occurs in the subarctic in any case. Since in these high latitude areas the snow lasts until spring melt, it will take longer to melt the deeper snowpack during the high sun period. It is becoming increasingly evident that spring snow has a major impact on the surface heat balance of the continents because of its high albedo. The magnitude of the scheme proposed in Quebec could have regional effects on the snow cover and when multiplied by the other inevitable large-scale river impoundments in northern Canada and Eurasia could have global impact. A counter argument is that there are so many lakes in northern latitudes already that a few more will not make any difference. This is a scientific controversy that will probably only be resolved after the fact. Furthermore the flow in the dammed rivers will be decreased in the spring when the reservoirs are being filled. In the case of James and Hudson Bays this may lead to a lag in the midsummer removal of ice from the eastern coastal areas, thus maintaining a high albedo for a longer period during the high sun season and again reducing the absorption of solar radiation and affecting the energy balance of fairly large areas. On a more general scale, Schneider (1976) draws attention to the importance of the northern rivers. Since they help maintain Arctic waters at a low-salinity level, so that they freeze readily and feed the Arctic ice pack, the diversion of north-flowing rivers in order to irrigate the grain lands of Canada and the USSR could eliminate the pack ice and create a much different climate in the northern polar regions. Schneider notes than an open polar ocean could result in more moderate and possibly more snowy winters. Whether this would start another glaciation of northern Canada and Europe as a result of the increased snowfall remains unknown but Schneider sees it as a definite possibility.

4.4 FOREST REMOVAL

In the subarctic and boreal forest zones of the world, large-scale forest destruction is largely a result of fire, caused either by lightning or by human activity. Fire is also used for deliberate land modification in some subarctic zones of the USSR. Destruction of the coniferous trees is usually total and as one approaches the tree line in northern Canada the time needed for re-establishment of the open forest canopy is around eighty years (Maikawa and Kershaw, 1976). Thus forest destruction represents a surface modification which lasts a considerable period. Rouse (1976) reports that burning of the subarctic forest near the tree line in north-western Canada results in a 10 to 15 per cent reduction in summer time net radiation and a generally more arid environment, with drier and warmer soils and reduced evapotranspiration. This lasts for a period in excess of twenty-five years. Figure 4.4 shows the energy/moisture relationship for a number of surfaces in the Canadian north during the summer period. It is evident that burning of the subarctic forest creates a surface which has less available radiant energy and which is substantially drier than other surface types in the high latitudes (Rouse *et al.*, 1977). As long as this impact is limited to natural lightning-caused fires, then it can be viewed as a natural sequence that has been operating through post-glacial times. However, more man-caused burning could upset what is considered a healthy balance of forest renewal. Palynological evidence produced by Nichols (1975) suggests that subarctic forests in the vicinity of Great Bear Lake in Canada are in disequilibrium with present-day climate, having been nurtured during a period of more favourable climatic conditions. He postulates that destruction of this forest would result in its replacement by a tundra vegetation sequence. In the USSR, Kryichov (1968) presents a warning concerning the use of forest burning as a land-use tool for creating a more favourable agricultural environment in the subarctic by blackening the soil and melting some of the permafrost near the surface. Apparently one of the primary recolonizers during vegetative recovery is a white tufted cotton grass with a high albedo. This serves to reduce the net radiation and inhibit soil warming, so that instead of retreating, the permafrost table advances toward the surface, creating a tundra environment which Kryichov refers to as 'pyrogenic tundra'.

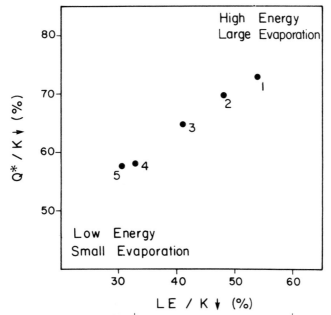

FIG. 4.4 SCALE OF AVAILABLE ENERGY ($Q*/K\!\downarrow$) AND RELATIVE DRYNESS ($LE/K\!\downarrow$) FOR A NUMBER OF HIGH LATITUDE SURFACES
(1) shallow tundra lake (2) tundra swampland (3) open subarctic woodland (4) dry upland tundra (5) burned subarctic woodland

In temperate and tropical areas forests have been transformed to other surface types on a large scale. It is estimated that from 18 to 20 per cent of continental areas have been changed from forests to grasslands or agricultural lands (SMIC, 1971). Indeed the large tropical savanna grasslands of Africa and South America appear to have been entirely man made, due to the burning of dry deciduous forests.

Forest removal results in substantial changes in the energy and water balances of a surface and in the availability of soil moisture. The SMIC (1971) report produced data on the energy balance changes after conversion from forest to agricultural land. These results are presented in a modified form in Table 4.2 and indicate a substantial reduction of net radiation and a large decrease in evapotranspiration in the drier agricultural lands and grasslands. Thus, as in the case of the burning of subarctic forests, the general trend is toward less available radiant energy Q^* and somewhat greater aridity.

The removal of forests in the wet tropics appears to create irreversible changes in the soils. Essential nutrients are maintained in the humus layer provided by the forest. Attempts to convert these areas into agricultural lands destroys the humus layer and transforms the soil into an impervious laterite with a small soil moisture storage capacity. This effectively prevents the re-establishment of tree growth. Thus forest destruction in both the subarctic and rainy tropics can be irreversible, though for quite different reasons. Newell (1971), in speculating on the effects that large scale deforestation of the Amazon basin could have on the atmospheric water balance, raises the possibility that the reduced water cycling between the surface and atmosphere would decrease the amount of latent heat input into the higher regions of the atmosphere. This could reduce the free convective turbulence in the tropical atmosphere and create an irreversible climatic change. Newell cautions, however, that the surface–atmospheric linkages are not well enough known to substantiate this hypothesis.

4.5 CLIMATE AND SPREADING DESERTS

The term 'desertification' became well used in 1972 with the dramatic spread of the Sahara desert southward into a formerly semi-arid zone known as the Sahel. This was

Table 4.2

ALBEDO AND ANNUAL HEAT BUDGET OF FORESTS, AGRICULTURAL LANDS AND GRASSLANDS

	α	Q^*	H	LE
Coniferous forest	12	100	33	67
Deciduous forest	18	88	22	66
Arable land (wet)	20	83	13	70
Arable land (dry)	20	83	25	58
Grassland	20	83	33	50

α albedo in per cent
Q^* net radiation
H sensible heat flux
LE latent heat flux
Energy balance data are presented in terms of 100 per cent for Q^* over a coniferous forest

Source: SMIC (1971) report, p. 173

accompanied by human starvation and migration on a major scale. Droughts in the semi-arid zone are not new but the extent, severity, and longevity of the Sahelian drought indicated clearly that misuse of the land was creating an impact which accentuated natural climatic variation.

Population pressures have resulted in overgrazing in the dry lands. This in turn has led to breakdown of the delicate biological balance in the soils, severe erosion and loss of structure of the topsoils, and a loss in their moisture retention capacity. This in turn inhibits the regrowth of the native vegetation and is, it is feared, an irreversible process. The loss of vegetation increases the albedo of the earth's surface, thus changing the surface radiation balance. Schneider (1976) points out a dramatic example of this. An earth resources satellite photograph of part of the Middle East showed a sharp dividing line between a dark area to the east which occupied the Israeli Negev Desert and a much lighter zone to the west, which until the 1967 war was part of the Egyptian Sinai Desert. The explanation is that the goats of the Bedouin nomadic herders defoliated so much land on the Egyptian side that the political boundary was visible from space in the form of an increased albedo. In similar fashion a 1973 satellite photo of the drought-stricken Sahel revealed a darker pentagonal shaped area within the encroaching desert. This proved to be a large ranch surrounded by a fence which kept out the herds of the nomadic tribesmen.

A number of intriguing and controversial theories relate the perpetuation of man-made deserts to local climatic change. Bryson (1967) argues that in the Rajasthan Desert of northwest India, overgrazing by goat herds has allowed erosion to place large amounts of wind-borne dust into the atmosphere. This dust increases the albedo of the earth-atmospheric system so that the greater loss of sunlight to space leads to atmospheric cooling. This creates a stable atmosphere in which the cooler air subsides. Such stable air in turn does not have the turbulent mechanisms for condensation and precipitation. Bryson also feels that air-borne dust has accentuated the Sahelian drought (Schneider, 1976). Charney *et al.* (1975) produce a different theory based on the fact that dry, sandy, unvegetated areas of the Sahara desert have a high albedo compared to much lower values for vegetated surfaces. In terms of equations (4.1) and (4.2) Q^* is small. In fact, for the earth-atmosphere as a whole, the annual radiation budget of the Sahara is close to zero and may even be negative. If overgrazing in the Sahel destroys the vegetation cover, then in like fashion the increased albedo will reduce its radiation balance. This favours a more stable atmosphere. Since the Sahel lies on the southern fringes of the Sahara where the atmospheric motion is dominantly subsiding and hence stable, the atmospheric circulation becomes tied in with that of the desert proper, and monsoon rain systems created by the migration of the Intertropical Convergence Zone are forced further towards the equator.

At this stage one must conclude that although there are no unequivocable studies on the role of man in enhancing the spread of deserts, his misuse of land is having a substantial climatic impact. It is a topic on which we can expect to hear more in the near future.

4.6 IMPACT OF URBANIZATION AND INDUSTRIALIZATION

A city generates its own distinctive climate. The important differences in comparison to its rural surroundings are summarized in Table 4.3. The increase in atmospheric particulates

Table 4.3

CLIMATIC DIFFERENCES BETWEEN URBAN AND RURAL AREAS

	Comparison with rural environment	
Element	More	Less
Dust particles	10 times	
Gases	5–25 times	
Precipitation	5–10%	
Temperature	0·5–1·0°C	
Solar radiation		15–20%
Albedo		50–100%
Wind speed		20–30%

Source: Landsberg (1962)

results from smoke and other combustion products, largely from industrial and home heating activities. Such particulates often become concentrated in a climatological or pollution dome which stretches upward to about a kilometre in height and from which they diffuse downwind. This urban particulate concentration is responsible for absorbing or reflecting up to 20 per cent of the solar radiation which arrives above the pollution dome and can also reduce solar radiation over the downwind countryside. Heavy particulate concentrations also serve to decrease visibility in the city and its immediate downwind surroundings and to increase the incidence of fog by up to 100 per cent. The latter occurs because water vapour readily condenses on hygroscopic atmospheric particulates to form small water droplets.

Almost all types of gaseous emission can emanate from cities but the most important are carbon dioxide, nitrogen dioxide, and reactive hydrocarbons. Along with particulates, carbon dioxide can have an impact on global climates and this will be discussed later. Nitrogen dioxide and the reactive hydrocarbons lead to a phenomenon known as photochemical smog, which is of world-wide concern in urban communities. The process is complex and not fully known, but in general, nitrogen dioxide absorbs ultraviolet light from the sun (part of $K\!\downarrow$) and is broken down into nitric oxide and atomic oxygen. The atomic oxygen combines with molecular oxygen to form ozone. Along with a number of other reactions, the ozone accumulates and reacts with hydrocarbons to form a group of secondary pollutants such as formaldehyde which are rather unpleasant. The most important effects of the photochemical smog on climatic elements are a reduction in visibility, the affect on solar radiation and an influence on precipitation in the form of fog (SMIC, 1971).

The increase in precipitation in a city and its surroundings is due to increased water vapour input from combustion sources, intensified thermal convection due to higher temperatures, increased mechanical convection due to the greater surface roughness of a city and larger concentrations of condensation nuclei due to the particulate output (Peterson, 1969). A small increase in precipitation has been found for cities in the mid-west of the United States (Peterson, 1969) but this does not usually exceed 10 per cent in comparison to rural surroundings. The larger amounts of precipitation are often accompanied by greater thunderstorm activity. Urban–rural comparisons are often difficult in the case of precipitation, since the urban influence can extend well downwind. A classical study of this effect has been provided by Changnon (1968) for the city of La

Porte, Indiana. La Porte lies about 50 km east of Chicago and 18 km southeast of the southern end of Lake Michigan. Since 1925 it has experienced an anomalous 30 to 40 per cent increase in total precipitation and in the number of days with rain, with thunderstorms, and with hail. The change is also closely related to the production curve of the Chicago iron–steel industrial complex. Changnon attributes the precipitation increases to the output of particulates, heat, and water vapour from the steel plants and related industries of Chicago. Subsequent studies by Changnon have indicated similar but less pronounced effects downwind from other cities.

The higher temperatures of cities are generally referred to as the 'urban heat island'. The effects are most notable at night when contrasts with rural surroundings are commonly 6°C in large cities and can rise as high as 12°C. Temperature anomalies are commonly associated with urban morphology. For example, a study by Chandler (1965) in London, England found that, under spring conditions of clear skies and light winds, highest temperatures correlated with the central business district. There was a slow decrease in temperature outward from the urban centre and a sharp temperature drop at the city boundaries. The overall temperature decrease was in excess of 6°C. The magnitude of the urban heat island shows a clear relationship to city size and to wind speed. The larger the city the stronger the winds which are necessary to dissipate the heat island effect (Oke and Hannell, 1970). The effects of the urban heat island do not normally extend beyond 300 m in the vertical plane, and above the heat island, temperatures can be lower than the surroundings.

The wind field within a city responds to the heat island. The higher temperatures in the city centre lead to lower pressure so that the winds converge from the surroundings to the centre where they rise and flow outward at higher levels to complete a cycle referred to as the urban wind cell. The wind cell reaches its maximum vigour at night and minimum by day, reflecting control by the intensity of the heat island.

Closely connected with the particulate and gaseous emissions, which on a local scale are concentrated in the climatological dome of cities, is the increase in particulates and carbon dioxide in the global atmosphere. A change in the concentration of particulates in the atmosphere may change both the global albedo and the absorption of solar radiation in the atmosphere. The interaction between particulates and sunlight is complex but most evidence suggests that an increase in the number of atmospheric particles would tend to cool the planet (SMIC, 1971) by increasing the planetary albedo. Since all cloud water droplets are formed by the condensation of water vapour on a particle, the change in particulate numbers in the atmosphere might be expected to affect the cloudiness. Indeed the principle in artificial rain making is to introduce appropriate extra condensation nuclei into a humid atmosphere in order to stimulate cloud formation. Similarly, the La Porte precipitation anomaly is believed to be due, at least in part, to the extra condensation nuclei produced by Chicago. However, on a global scale there is no indication that man's activities have affected the amount of cloudiness. It appears that there are more than enough condensation nuclei in the atmosphere from natural sources so that the extra industrial output has no effect (SMIC, 1971).

Most recent evidence points to the rapid increase in atmospheric carbon dioxide as a factor most likely to cause climatic modification and perhaps major climatic change. As a result this is a feature worth considering in some detail. Along with water vapour, carbon dioxide is the most important gas in absorbing long wave radiation leaving the earth's surface and directing much of it back to the surface to create the 'greenhouse effect'. Therefore, the greater the concentration of carbon dioxide in the atmosphere the warmer the surface of the earth. As a result of the burning of fossil fuels the atmospheric concentration of carbon dioxide has increased from 290 parts per million (ppm) before the industrial revolution to 320 ppm today and, at the present rate of increasing fuel consumption, will rise to around 400 ppm by the beginning of the twenty-first century and to 600 ppm somewhere between 2025 and 2040 (Schneider, 1976). There have been a number of attempts to model the effect of a doubling of carbon dioxide concentration and they all indicate an increase in global temperatures, but the magnitude of the estimated increases varies from 0·7 to 9·6°C. In an effort to resolve such a wide variation, Schneider (1975) has compared the different modelling attempts. He concludes that the best estimate is that a doubling of CO_2 content would raise average global temperature by 1·5 to 3·0°C and that this estimate is as likely to be too low as too high.

The effects of an increase in global temperatures of the order of 2°C would be many. Manabe and Weatherald (1975) and Schneider (1975) note that the temperature increase in the high latitudes would be much greater than in the tropics. This is due to a decrease in albedo as the snow and sea-ice cover recede and to a lack of atmospheric mixing in the polar latitudes. A small warming of the Arctic by a few degrees would start a melting of the ice pack which covers 90 per cent of the Arctic Ocean. With melting there would be a large decrease in albedo which would cause additional solar radiation to be absorbed, thereby hastening the warming trend until the ice disappeared. In the northern hemisphere, melting of the sea ice would not affect world ocean levels to any degree. However, this might not be the case in the southern hemisphere. Mercer (1978) raises the interesting fact that Antarctic ice cap consists of two unequal parts with quite different characteristics. The larger East Antarctic ice sheet is land-based and has been long-established. The smaller and younger West Antarctic ice sheet is marine based and represents a vertical displacement of ocean water. It is held to the more stable land-based ice sheet by the Ross Ice Shelf and Ronne Ice Shelf and needs, Mercer calculates, summers which are up to 10°C colder than the East Antarctic ice cap in order to survive. If summer temperatures rose by 5°C, the Ross and Ronne ice shelves would melt and the West Antarctic ice sheet would be undermined and rather drastic deglaciation in the West Antarctic would probably commence. Complete melting of the West Antarctic ice sheet would lead to an average 5 m rise in world sea levels. Although at first glance such speculations may appear to lie in the realm of science fiction, they have very real possibilities in a climatically unstable world and have many serious scientific investigators both intrigued and concerned.

4.7 CONCLUSION

There is a wide range in man's modification of climate from relatively local effects to global impacts. Changes in the terrestrial surface modify the surface energy balance which in turn can influence the atmosphere above. Such changes include large scale river diversions and irrigation; forest removal for agriculture and due to fire; vegetation denudation caused by overgrazing and overpopulation in sub-desert areas and the urbanization and industrialization of the landscape. Some of these processes may be irreversible, such as forest removal in subarctic and tropical zones and possibly the destruction of the soil water-holding ability in semi-arid and subhumid regions. Urban and industrial expansion has an impact far beyond the land area in which it occurs, due to the emission of particulates and carbon dioxide into the atmosphere. It appears almost certain that, if the carbon dioxide build-up in the atmosphere continues unabated, it will lead to an overall warming of planet earth and that this will be most strongly felt in high latitudes. The melting of the arctic ice pack and some continental ice sheets lies in the realm of possibility.

REFERENCES

BROWN, D. M., McKAY, G. A. and CHAPMAN, L. J., 1968, 'The climate of southern Ontario.' *Climatol. Studies*, 5, Atmosph. Environ. Serv., Toronto.

BRYSON, R. A., 1967, 'Possibilities of major climatic modification and their implications: Northwest India.' *Bull. Amer. Met. Soc.*, 48, pp. 136–42.

BUCKLER, S. J., 1973, 'The temperature and evaporation regime on Lake Diefenbaker during 1972.' *Hydromet. Rep. 8*, Atmos. Environ. Serv., Regina Airport, Saskatchewan.

BUTORIN, N. V., VENDROV, S. L., DYAKONOV, K. N., RETEYUM, A. Y. and ROMANENKO, V. I., 1973, 'Effect of the Rybinsk Reservoir on the surrounding area', *Man-Made Lakes: Their Problems and Environmental Effects*, Geophys. Monog., 17, Amer. Geophys. Union, pp. 246–50.

CHANDLER, T. J., 1965, *The Climate of London* (Hutchinson, London).

CHANGNON, S. A., 1968, 'The La Porte weather anomaly — fact or fiction.' *Bull Amer. Met. Soc.*, 49, pp. 4–11.

CHARNEY, J., STONE, P. H. and QUIRK, W. J., 1975, 'Drought in the Sahara: a biophysical feedback mechanism.' *Science*, 187, pp. 434–5.

CORK, H. E., 1974, 'Lake Diefenbaker energy budget — summer 1973.' *Hydromet. Rep., 11*, Atmos. Environ. Serv., Regina Airport, Saskatchewan.

DEKKER, G., 1972, 'A note on the Nile.' *Water Resources Res.*, 8, pp. 818–28.

FELS, E. and KELLER, R., 1973, 'World register on man-made lakes', *Man-Made Lakes: Their Problems and Environmental Effects*, Geophys. Monog. 17, Amer. Geophys. Union, pp. 43–9.

FLOHN, H., 1971, 'Investigation on the climatic conditions of the advancement of the Tunisian Sahara.' *World Met. Organization Tech. Note*, 116.

HARBECK, G. E., KOHLER, M. A. and KOBERG, G. E., 1958, 'Water loss investigations: Lake Mead studies.' *US Geol. Surv. Prof. Paper*, 298.

KORNIENKO, V. I., 1969, 'The effect of evaporation from Lake Baikal on precipitation in adjacent regions.' *Leningrad GL. Geofiz Observ. Tr.*, 247, pp. 122–6.

KRYICHOV, V. V., 1968, 'Soils of the far north should be conserved.' *Piroda*, 12, pp. 72–4 (in Russian). Translated in *The Effect of Disturbance on Permafrost Terrain*, US Army CRREL, SR, 138.

LANDSBERG, H. E., 1962, 'City air — better or worse', *Symposium: Air Over Cities*, (US Public Health Service, Taft Sanitary Eng. Center, Cincinatti, Ohio), Tech. Rept., A62–5, pp. 1–22.

LVOVICH, M. I., 1969, *Vodnie Resoursi Buduschevo* (Water Resources of the Future) (in Russian, *Prosreschenie ed., Moscow)*. Quoted in the SMIC (1971) report.

MAIKAWA, E. and KERSHAW, K. A., 1976, 'Studies on lichen-dominated systems. 19. The post-fire recovery sequence of black spruce-lichen woodland in the Abitau Lake Region, NWT.' *Can. Jnl. Bot.*, 54, pp. 2679–87.

MANABE, S. and WEATHERALD, R. T., 1975, 'The effects of doubling the CO_2 concentration on the climate of a general circulation model.' *Jnl. Atmos. Sci.*, 32, pp. 3–15.

McVehil, G. E. and Peace, R. L., 1965, 'Some studies of lake effect snowfall from Lake Erie.' *Gt. Lakes Res. Div., Univ. Mich. Publ.*, 13, pp. 262–72.

Mercer, J. H., 1978, 'West Antarctic ice sheet and CO_2 greenhouse effect: a threat of disaster.' *Nature*, 271, pp. 321–5.

Nemec, J., 1973, 'Interaction between reservoirs and the atmosphere and its hydrometeorology elements', *Man-Made Lakes: Their Problems and Environmental Effects*, Geophys. Monog., 17, Amer. Geophys. Union, pp. 398–405.

Newell, R. E., 1971, 'The Amazon forest and atmospheric general circulation', *Man's Impact on the Climate*, ed. W. H. Mathews, W. W. Kellogg and G. D. Robinson (MIT Press, Cambridge, Mass.), pp. 457–9.

Nichols, H., 1975, 'Palynological and paleoclimatic study of the boreal forest-tundra ecotone in NWT.' *INSTAR, Colorado, Occas. Pap.*, 15.

Oke, T. R. and Hannell, F. G., 1970, 'The form of the urban heat island in Hamilton, Canada.' *World Met. Org. Tech. Note*, 108, pp. 113–26.

Peterson, J. T., 1969, *The Climate of Cities: Survey of Recent Literature*, National Air Pollution Control Administration, AP–59 (US Dept. Health, Education and Welfare, Raleigh, N. Carolina).

Rouse, W. R., 1976, 'Microclimatic changes accompanying burning in subarctic lichen woodland.' *Arctic and Alpine Res.*, 8, pp. 357–76.

Rouse, W. R., Mills, P. F. and Stewart, R. B., 1977, 'Evaporation in high latitudes.' *Water Resources Res.*, 13, pp. 909–14.

Schneider, S. H., 1975, 'On the carbon dioxide–climate confusion.' *Jnl. Atmos. Sci.*, 32, pp. 2060–6.

1976, *The Genesis Strategy* (Plenum Press, N.Y.).

Schneider, S. H. and Dickinson, R. E., 1974, 'Climatic modelling.' *Rev. of Geophys. and Space Phys.*, 12, p. 447.

SMIC, 1971, *Report of the Study of Man's Impact on Climate* (MIT Press, Cambridge, Mass. and London).

Hydrosphere

5

Hydrological Processes

D. E. WALLING
University of Exeter

Hydrology is no longer limited to the geophysical science of the natural circulation and storage of waters on the earth. While that circulation is still the dominant feature, it is increasingly modified and managed by man, and the hydrologist who does not recognise this will be left back in the nineteenth century!

W. C. Ackermann (1969)

5.1 INTRODUCTION

These sentiments were the keynote of an introductory address delivered to an international hydrology seminar which took place in Urbana, Illinois in 1969. Faced with the task of highlighting recent trends within the science of hydrology, Ackermann emphasized the need for a greater awareness of the impact of man on hydrological processes and added further support to a view which had become increasingly voiced during the 1960s. In response to such pressure, the theme of 'the influence of human activity on hydrological regimes' had been included as one of the principle foci of the International Hydrological Decade (IHD) (1965-74). This period of international co-operation in hydrology, launched by UNESCO, has been succeeded by the International Hydrological Programme (IHP) and two of the eight major projects of the first phase of this programme again reflect mounting interest in and concern for the impact of man on the hydrological cycle.

Numerous studies, working groups, and symposia resulted from this international activity, and the vast body of evidence and knowledge concerning man's impact on hydrological processes and associated problems that has accumulated, reflect a widespread recognition of the view echoed by Ackermann. In some instances these effects have been largely *intentional*, as in the case of river regulation schemes and exploitation of groundwater for water supply, but more widespread are the *inadvertent* effects whereby, for example, forest clearance has markedly changed the flow regime of a river basin or agricultural activity has influenced the water quality of local streams. In both cases a variety of process mechanisms are involved and the hydrological modifications may range from the almost imperceptible and perhaps insignificant to the catastrophic. However, assessing human impact itself poses philosophical problems, for there are few remaining areas of the contemporary world where true wilderness conditions exist to provide a yardstick against which to evaluate change. In the British Isles, for example, any area selected as a natural baseline is likely to have been subject to human interference over the past 3000 years or more.

Despite a general recognition of the far-ranging significance of man's impact on the hydrological cycle, few hydrology texts have explicitly covered this aspect of the subject. This chapter attempts to review some of the evidence now available and to provide an insight into the nature and extent of this impact.

5.2 PROCESS MODIFICATIONS

Figure 5.1 presents a simplified representation of the hydrological cycle operating in an essentially natural drainage basin. Inputs of precipitation are distributed through a number of stores by a series of transfer processes and are output as channel flow, evapotranspiration and deep leakage. Thus, in simple terms, water entering a drainage basin is intercepted by vegetation and subsequently evaporates or reaches the ground surface as throughfall and stemflow. In the absence of vegetation, the precipitation reaches the ground directly. At the ground surface water infiltrates into the soil or is retained in surface storage. The surface water may move downslope as surface runoff or be slowly evaporated. Moisture held within the soil is subject to surface evaporation and plant transpiration, downward percolation to the water table and the groundwater body, and downslope movement as throughflow and interflow. Water held within the ground-water body is similarly subject to losses by plant transpiration and evaporation when the water table is close to the surface, upward movement into the unsaturated soil by capillary action, deep leakage to adjoining groundwater bodies, and slow surface outflow into springs, seeps, and river channels as baseflow. The river channel receives variable contributions of surface runoff, throughflow and interflow, and baseflow which together contribute to the time–variant outflow hydrograph of the drainage basin.

An attempt has been made to point to the various ways in which this cycle could be modified through man's activities by considering first, modification of internal processes and secondly, additional moisture inputs (Fig. 5.1). For convenience, process modifications have been taken to include both the transfer processes and the functioning of the various stores and a glance at the appended list of significant human activities clearly demonstrates that in most locations the hydrological cycle is far from a natural system. Specific activities will not in all cases produce consistent effects. For example, whereas in many areas cultivation has been shown to reduce surface infiltration and increase surface runoff there is a significant body of evidence from certain areas of the USSR indicating that autumn ploughing can cause a decrease in surface runoff associated with spring snowmelt (e.g. Shevchenko, 1962). Additional moisture inputs may further modify the system and

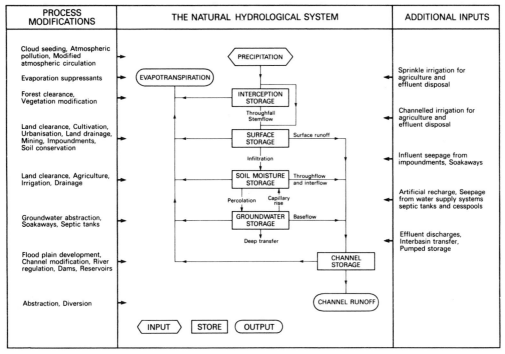

FIG. 5.1 MAN AND THE NATURAL HYDROLOGICAL SYSTEM
A simplified diagrammatic representation of the hydrological processes operating in a drainage basin and some examples of human interference associated with process modifications and additional moisture inputs

a number of potential sources have been listed in Figure 5.1 and related to individual flow pathways. Although extreme, it would not be difficult to find situations where inputs of irrigation water exceed the natural precipitation input to a basin, or where inputs to channel storage from effluent outflow sometimes exceed natural runoff.

Some specific information concerning the nature and extent of process modifications is provided by Figure 5.2, where a few of the many possible examples have been illustrated. In the case of evapotranspiration, some of the results of a study by Gleason (1952) aimed at estimating the losses occurring under different land uses within the Raymond Basin, Pasadena, California, are cited (Fig. 5.2A). If the areas of lawn and trees are taken to be the most representative of natural conditions, then the extreme case of change to pavement and streets can be seen to cause an 86 per cent reduction in evapotranspiration. Instances of increased losses could be found in other urban areas where sprinkler irrigation is widely used on domestic lawns and recreational grassland.

An indication of the impact of agricultural land use on infiltration rates is afforded by Figure 5.2B. The cumulative infiltration graphs reported by Holtan and Kirkpatrick

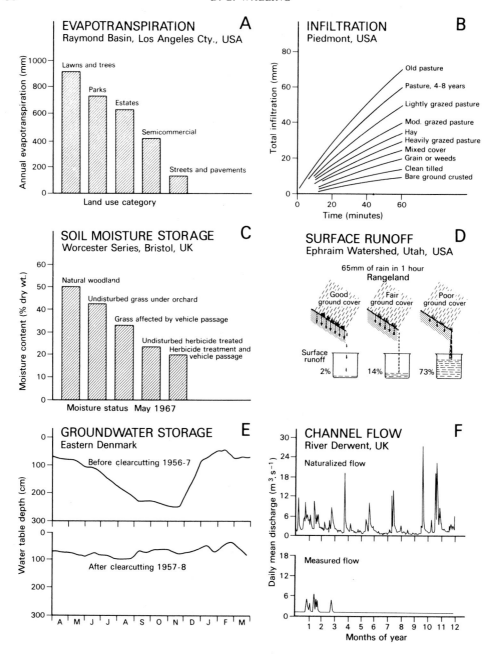

FIG. 5.2 SOME EXAMPLES OF MAN'S IMPACT ON INDIVIDUAL HYDROLOGICAL PROCESSES.
(Sources: (A) data in Gleason, 1952; (B) Holtan and Kirkpatrick, 1950; (C) data in Hills, 1969; (D) Noble, 1965; (E) Holstener-Jorgensen, 1967; and (F) Richards and Wood, 1977)

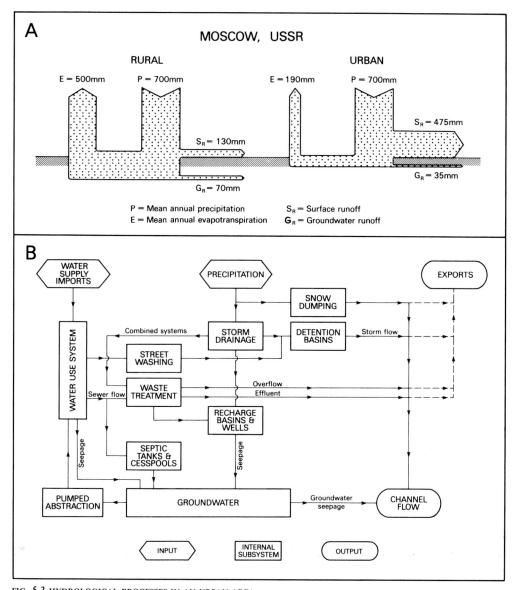

FIG. 5.3 HYDROLOGICAL PROCESSES IN AN URBAN AREA
(A) compares the water balance of a rural area and a heavily built-up zone of Moscow, USSR. (Source: data presented by Lvovich and Chernogaeva, 1977). (B) illustrates some of the additional components associated with the functioning of an urbanized catchment

(1950) from measurements on Durham, Madison and Cecil soils in South Carolina similarly lack a natural baseline for comparative analysis, but if the response of old pasture is used for this purpose, reductions in infiltration rates of up to 80 per cent or more are seen to result from various cultivation and grazing practices. Reduced infiltration in turn leads to decreased soil moisture storage and a measure of the potential reduction can be found in the work of Hills (1969) who studied the moisture content of soils under different management practices in the area around Bristol, UK. Measurements made on the heavy soils of the Worcester Series (Fig. 5.2C) demonstrate decreases in moisture storage by more than 50 per cent in response to the passage of agricultural vehicles and herbicide treatment. These changes were related primarily to reduced infiltration, although the decrease in available storage space caused by compaction could also prove significant in many areas. Infiltration rates clearly exert a critical control over the generation of surface runoff by controlling the relative volumes of water remaining at the surface and passing down through the soil. An example of the impact of grazing activity on infiltration and more particularly surface runoff is provided by Figure 5.2D. Based on the work of Noble (1965) in the Mount Ephraim area of intermontane Utah, this demonstrates how a reduction in ground cover from good (60–75 per cent cover) to poor (10 per cent), as a result of overgrazing, increases the proportion of rainfall appearing as surface runoff from 2 per cent to as much as 73 per cent, an increase of over thirty-five-fold.

Groundwater storage will respond to changes in infiltration and recharge and also to modified evapotranspiration rates where the water table is sufficiently close to the surface to render loss by capillary rise or plant uptake significant. The work of Holstener–Jorgensen (1967) at the Danish Forest Experimental Station illustrated in Figure 5.2E provides an example by demonstrating the effects of clearing a 75-year-old beech stand on groundwater levels. Before clearance, the water table exhibited a marked seasonal variation, dropping during the summer in response to water uptake through the tree roots, but this pattern almost completely disappeared after clearing.

Finally, Figure 5.2F, based on the work of Richards and Wood (1977), presents a graphic example of the effects of modifying channel storage, in this case by reservoir construction, on channel flow regime. The recorded flow in the River Derwent at Yorkshire Bridge, a site downstream from the Ladybower Reservoir, contrasts markedly with the reconstructed or naturalized record in showing an almost complete absence of peaks and a constant minimum flow throughout 85 per cent of the year.

Figure 5.2 illustrates some of the process modifications cited on Figure 5.1 and demonstrates that man can profoundly influence the individual components of the hydrological cycle. In addition, it is clear that such modifications are widespread occurrences, rather than isolated extremes. With the exception of Figure 5.2A, however, the examples are all taken from essentially rural locations and even greater modifications can be found in urban areas. Under tarmac and concrete, infiltration will approach zero and the effective 'roofing over' of the groundwater body will cause rapidly falling water table levels. As noted previously, evapotranspiration losses from buildings and paved areas will be drastically reduced, and the storm runoff dynamics of a system of roofs, gutters, roads, gullies, stormwater drains and culverts will differ radically from those of a natural surface.

Some indication of the change in the relative importance of the various components of the hydrological cycle associated with the transition from rural to urban conditions is provided by Figure 5.3A. This depicts simplified water budgets for rural and urban conditions in the area of Moscow, USSR calculated by Lvovich and Chernogaeva (1977). In that area, the establishment of urban conditions has resulted in a decrease in evapotranspiration and groundwater runoff by 62 per cent and 50 per cent respectively and an increase in total runoff by 155 per cent.

More detailed investigation of an urban hydrological system reveals numerous features going far beyond simple modification of the natural system shown in Figure 5.1. Figure 5.3B attempts to depict some of the major components of such a system under fully urbanized conditions and introduces the need to incorporate water import, waste treatment, artificial recharge, storm drainage, and effluent disposal components. In some instances a large proportion of the output may be diverted out of the basin to the sea or to another river system.

Modifications to individual processes will clearly have repercussions throughout the hydrological system and for this reason they are rarely studied in isolation. In most investigations of the impact of man, the drainage basin is employed as the unit of study so as to include the complete system. Streamflow output from the basin has often been used as an easily measured yet sensitive parameter integrating the various internal processes. Thus, it has often proved easier to gauge the precipitation input and runoff output of a catchment over a period of time and thereby deduce process changes, than to undertake detailed instrumentation of individual components, although such measurements can be extremely valuable in more rigorous work. Further discussion of man's impact on hydrological processes will therefore focus on results that have been obtained from catchment studies and experiments.

5.3 CATCHMENT STUDIES AND EXPERIMENTS

The use of catchment studies to assess the impact of human activity on hydrological response can be traced back to the 1890s and work in Switzerland (Engler, 1919) and to the classic Wagon Wheel Gap investigation initiated in Colorado, USA in 1910 (Bates and Henry, 1928). The former study involved the instrumentation of two small watersheds to evaluate the contrasts in response resulting from differences in land use and was essentially comparative in approach. The Wagon Wheel Gap study can, however, be classed formally as an experiment. It involved the comparison of two small forested basins for an initial period in order to derive a *calibration* and the subsequent clearance of one of them. The effects of clearance were assessed using the initial calibration to provide estimates of departure from natural response.

Similar studies were subsequently undertaken at many other research stations in the USA, for example at Coweeta in North Carolina (Dils, 1957), and in other countries (Pereira, 1973; Rodda, 1976), but it was the Representative and Experimental Basin Programme of the IHD that provided the stimulus for a major upsurge in catchment studies. Whilst representative basins were established primarily to provide data representative of a hydrological region, experimental basins were set up to study the effects of cultural changes on drainage basin dynamics and explicitly involved deliberate modification of the catchment characteristics. The numerous experimental catchment investigations that were carried out within the framework of the IHD provide a valuable source of information on man's impact on basin response for many areas of the world.

The need to provide a baseline or calibration against which to assess changes in catchment behaviour has generated a number of strategies for experimental catchment research. In the first, the classic paired or control watershed study, two essentially similar basins are instrumented and calibrated against each other. The effects of subsequent modifications to one of them can be evaluated using this calibration to estimate the natural response of the modified basin from the records of the unaltered control. Figure 5.4A presents an example of this type of watershed calibration. It relates to a paired watershed experiment at the Coweeta Hydrologic Laboratory in North Carolina aimed at assessing the impact of clear-felling on runoff yield. The three-year calibration period was used to derive a relationship between the monthly water yields of the two catchments and the increases in yield consequent upon felling were estimated using this relationship. The felling was repeated twenty-three years later and the almost identical increase in water yield demonstrates the consistency of the experiment.

Where it is difficult to find suitable twin watersheds, the single watershed approach has often been adopted. In this a single catchment is calibrated in its natural state against hydrometeorological variables, such as precipitation input, which will not be modified by the subsequent alterations to the basin. Figure 5.4B illustrates the use of this approach in the study of the impact of building activity on the storm runoff response of a small catchment on the margins of Exeter (Walling and Gregory, 1970). A multiple regression equation relating storm runoff volume to precipitation character and antecedent moisture status was fitted to the runoff record for the period of undisturbed condition and this was used to estimate the natural response to storm rainfall after building activity occurred. By comparing these values with the recorded runoff volumes, the magnitude of the increase can be assessed. In this particular case, disturbance of 25 per cent of the surface area of the basin resulted in an average increase in storm runoff volumes of 2·4 times.

Where a number of research catchments are located within a small area it is possible to devise experiments to evaluate the effects of a number of different watershed treatments simultaneously. This strategy is often termed the multiple watershed approach and a good example is provided by work undertaken at the Moutere Soil Conservation Station in South Island, New Zealand by the New Zealand Ministry of Works (1968). Aimed at studying the effect of different land management practices, this study involved twelve small basins (Fig. 5.4C). From 1962 to 1965 the catchments were calibrated and a programme of treatment was subsequently initiated (Table 5.1). Some of the results obtained are presented in Figure 5.4C(i). Here the influence of gorse clearance and subsequent arable cultivation on storm peak discharges from catchment 10 has been evaluated using catchment 5 as a control. Increases in storm peak discharge in excess of one order of magnitude are apparent in the smaller events. In an experiment of this type treatments can be changed once sufficient data have been collected for a particular management practice. In 1970 a further programme of cultural changes was initiated at Moutere (New Zealand Ministry of Works, 1971), and this included the planting of exotic forest (*Pinus radiata*) on catchments 8, 13, and 14. Peak runoff distributions for the years 1975 and 1976, averaged for three catchments in pasture (2, 5, 15) and the three planted with pines, are presented in Fig. 5.4C(ii). Catchments in pines yielded fewer peaks in all magnitude classes.

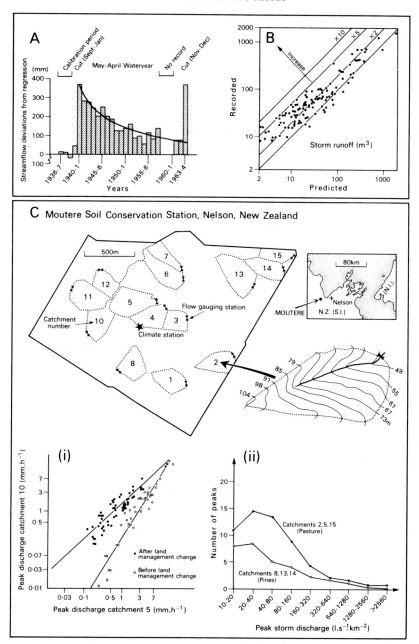

FIG. 5.4 CATCHMENT STUDIES
(A) illustrates the results of a paired watershed study of the effects of forest clearance on runoff. (Source: Hibbert, 1967). (B) shows how data from a single watershed study have been used to assess the increase in storm runoff volumes associated with urban development. (C) provides a plan of the multiple watershed study in progress at Moutere, New Zealand and presents some results relating to the effects on peak flows of conversion of a small gorse catchment to cultivation and cropping (i) and of forest planting (ii). (Sources: New Zealand Ministry of Works, 1968 and 1978; and Scarf, 1970).

Table 5.1

Treatment	Catchment numbers
1 Mob stocking[1] plus contouring	2 and 6
2 Set stocking[2] plus contouring	4 and 15
3 Mob stocking no contouring	5 and 14
4 Set stocking no contouring	1 and 3
5 Gorse	8 and 13
6 Cultivated and sown into crops	10 and 12

[1]Intermittent grazing, high animal density
[2]Continuous grazing, low animal density

Source: New Zealand Ministry of Works (1968)

5.4 EVIDENCE FROM CATCHMENT STUDIES

The focusing of attention through the IHD on man's impact on the hydrological cycle and on catchment studies as a means of assessing this impact has produced a wealth of empirical evidence concerning the effects of human activity on basin response. It would prove a daunting task to review such evidence within this short chapter and, instead, a number of case studies can be considered, highlighting particular activities.

5.4a Land clearance

Over large areas of the globe, the development of agriculture, both by early civilisations and by colonists opening up the New World, must have produced far reaching changes in drainage basin response. Clearance continues today for extension of agriculture and for timber production and a measure of the potential magnitude of these changes in provided in Figure 5.5A. This presents the results of a catchment experiment in the Allegheny Mountains of West Virginia, cited by Harrold (1971). The response of a clearcut catchment to a heavy summer storm is compared to that of an undisturbed control basin and an increase in flood magnitude of about six times is evident. This can be accounted for by reduced interception, by reduced infiltration, particularly as a result of surface compaction, and by a general reduction in evapotranspiration losses. Higher flood peaks are usually accompanied by increased total annual runoff and increases of more than 50 per cent have been reported for certain locations in the USA by Anderson et al. (1976) in their valuable review of the hydrological role of forests.

The effects of fire when used knowingly or accidentally as an agent of land clearance are particularly extreme. Figure 5.5B refers to the impact of a bushfire on a small watershed in the Entiat Experimental Forest in the Cascade Mountains of north-central Washington (Helvey, 1972). Although essentially a natural lightning-caused phenomenon, the findings are equally applicable to man-induced fires. The increases in flood magnitude and total runoff associated with destruction of vegetation and soil organic horizons are graphically illustrated in Figure 5.5B which compares the annual hydrograph for the year after the fire with the previous calibration record. The peak associated with spring snowmelt was about three times greater than those recorded previously and occurred about a month earlier as a result of increased heat absorption by the blackened timber.

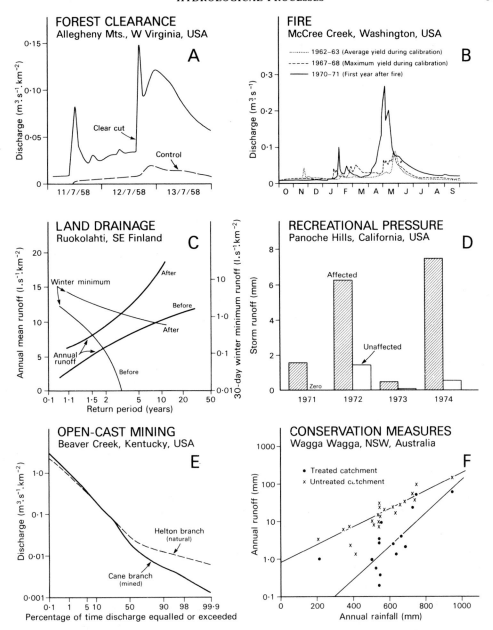

FIG. 5.5 MAN'S IMPACT ON CATCHMENT RESPONSE
Results of several case studies discussed in the text are illustrated. (Sources: (A) Harrold, 1971; (B) Helvey, 1972; (C) Mustonen and Seuna, 1971; (D) data in Snyder *et al.*, 1976; (E) Collier *et al.*, 1970; and (F) data in Adamson, 1974

5.4b   Land drainage

In many locations, the development of intensive agriculture and of commercial forests is accompanied by land drainage in order to lower the water table or, in the case of impermeable soils, to remove surface water. In several cases drainage activities have been linked to increased flooding (e.g. Howe *et al.*, 1967), but the evidence is somewhat contradictory. For example, McCubbin (1938) cites an analysis of the flood hydrographs of the Iowa and Des Moines Rivers in the USA before and after the installation of field drains over a large proportion of the catchment areas, in which no significant change was detected. Much depends on the complex interaction of groundwater levels, rainfall intensity, soil moisture storage and soil transmissibility and Rycroft and Massey (1975) present evidence which suggests that field drainage can actually reduce flood flows.

More convincing evidence of the impact of drainage on catchment response can be found in work undertaken in Finland on the effects of peat drainage for forestry. About one third of the land area of Finland (*c.* 10 million ha) is peatland and by 1970 about 3·5 million hectares of this had been drained. A catchment study initiated at Ruokolahti in 1935 (Mustonen and Seuna, 1971) has compared two basins, one of which was ditched and drained in the period 1958–60. In this environment, drainage causes lowering of the water table and reduced evapotranspiration and the annual mean runoff for the drained catchment increased by over 40 per cent. Figure 5.5C further demonstrates the effect of drainage in increasing flows by comparing the frequency curves of annual mean runoff and winter minimum runoff for the before and after periods.

5.4c   Recreational pressure

Whilst land clearance provides an example of man's inadvertent effects on the hydrological cycle, land drainage must be viewed as a more purposeful modification of those processes. Recreational activity provides a further example of an inadvertent effect, although in this case the physical impact on the landscape is essentially unintentional. Reference can be made to a study undertaken in the arid zone of central California (Snyder *et al.*, 1976). In this area recreational pressure had increased very rapidly and the use of off-road vehicles, principally motorcycles, was giving particular cause for concern in terms of damage to the fragile desert ecosystem. A hydrological study was initiated in a canyon area of the Panoche Hills where as many as 2000 motorcycles could be found during peak weekend periods. The closure of the area to the public in 1970 provided an opportunity to isolate two small watersheds: one that had been the centre of intensive hill climbing activity and another that was essentially undisturbed. Some of the results portrayed in Figure 5.5D indicate that, even during the recovery period, runoff from the affected area was drastically increased as a result of lack of vegetation cover and of soil compaction leading to the initiation of rill and pipe erosion on stressed areas.

5.4d   Strip mining

Strip mining provides an example of severe landscape disturbance and the related surface destruction, deep excavation, and in some cases pumping to lower water levels, can profoundly modify catchment response. A useful perspective on the hydrological effects of such activity is provided by work undertaken in the Beaver Creek Basin of Kentucky (Collier *et al.*, 1970). Two small tributary basins, one undisturbed and one with strip mining over 10·4 per cent of its area, were compared over the period 1957-66. The gross runoff from the two catchments was effectively similar, but the detailed distribution of that runoff differed significantly. The mined watershed evidenced increased peak flows and decreased minimum flows and this increased flow variability is reflected in the flow duration curves presented in Figure 5.5E. The increased peaks may be attributed to surface disturbance and compaction causing reduced infiltration and the reduction in low flows would in turn be conditioned by reduced infiltration and modified groundwater storage caused by deep excavations. It must, however, be recognised that increased flow variability may not always result from surface mining activity, particularly where extensive pumped drainage is involved. In the latter case, runoff volumes may increase and exhibit a more balanced flow regime with reduced flood peaks and increased minimum flows. This was the situation described by Golf (1967) from a catchment study in the brown coal mining area of Lower Lusatia, GDR, where lowering of the water table is essential for mining operations and where 6·3 m^3 of water are pumped for every tonne of coal mined.

5.4e   Conservation measures

For the most part, discussions of the hydrological effects of human activity emphasize the essential detrimental changes, but the potential impact of man's positive action in manipulating basin response to produce a more favourable regime must also be recognized. The planting of forests for flood abatement can reverse the trends discussed above in the context of land clearance, and conservation measures are widely employed as an essential part of land management. In many instances man may merely be repairing damage that he himself has caused but the element of positive control nevertheless remains. Out of numerous possible case studies an example drawn from catchment studies undertaken at Wagga Wagga, New South Wales, Australia, can be introduced. Here clearance of native woodland for pasture and crops resulted in increased runoff and accelerated erosion and a paired catchment experiment (Adamson, 1974) has been used to evaluate the long-term effects of soil conservation structures and land-use management on runoff and soil loss. These measures resulted in an average reduction in runoff of 74 per cent and the marked change in the annual rainfall/runoff relationship is clearly evident from Figure 5.5F.

5.4f   Urbanization

The extreme modification of hydrological processes resulting from urbanization has already been reviewed and the evidence available from catchment studies provides a similar perspective on this aspect of human activity. Increases in flood magnitude have been widely documented and Figure 5.6A illustrates an attempt by Leopold (1968) to synthesise the results of a number of catchment studies in the USA in terms of the effects of impervious area extent and area served by storm sewers on the mean annual flood from a 2·59 km^2 (1·0 mi^2) catchment. Increases in flood peaks of three to four times are associated with highly urbanized areas and may be ascribed partly to increased runoff volumes related to reduced infiltration and partly to the more efficient drainage network.

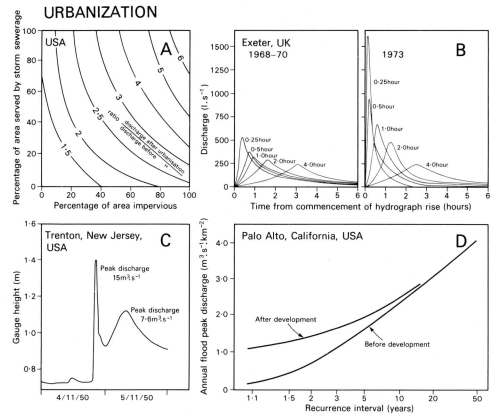

FIG. 5.6 THE EFFECTS OF URBANIZATION ON CATCHMENT RESPONSE
A general relationship between degree of urbanization of a catchment and the magnitude of the mean annual flood from a 2.59 km² basin, proposed by Leopold (1968), is presented in (A). The impact of urban development on hydrograph shape is shown in (B) by comparing unit hydrographs for natural and developed conditions. (C) demonstrates the effects of urbanization in the lower part of a 230 km² basin on storm hydrograph form, (Source: Waananen, 1969) and (D) presents hypothetical flood frequency curves for a small basin under natural and urbanized conditions. (Source: Crippen and Waananen, 1969)

Storm hydrographs from urban catchments characteristically exhibit sharper peaks and reduced lag times between rainfall and runoff as a result of this increased efficiency. To develop this point, Figure 5.6B portrays some results collected by the author from a small experimental basin on the margins of Exeter, UK, which has been used to evaluate the impact of suburban development on storm runoff response. Unit hydrographs, the hydrographs resulting from 1·0 cm of *effective* rainfall falling over the basin within a specified duration, have been derived from the records for the catchment in its natural state and for 1973 when about 25 per cent of the area was being developed. Although relating primarily to the construction phase, the contrasts between the unit hydrographs for these two periods clearly reflect changes in the runoff concentration and transmission properties of the basin. The volumes associated with the hydrographs for the two periods are identical (equivalent to 1·0 cm of effective rainfall) and the contrasts in peak flow result specifically from changes in hydrograph shape. The expected decrease in lag times is also evident in 1973.

Most of the evidence concerning the impact of urbanization on basin response relates primarily to small catchments where development covers a sizeable proportion of the surface. Changes in flood runoff may be obscured within a larger catchment as they are routed downstream and combined with the output from more natural areas. Changes in hydrograph timing can, however, be very important in this context, for Waananen (1969) cites the example of the 230 km^2 Assunpink Creek basin in New Jersey, USA, where the presence of an urban area in the lower reaches results in a storm hydrograph with a sharp peak preceding the flatter, longer duration peak produced by the undeveloped portion (Fig. 5.6C). The volume of runoff involved in the urban-derived peak is relatively small but the peak discharge is nearly double that of the main hydrograph.

Also of significance to attempts to assess the impact of urbanization, or indeed any other human activity, on basin response are considerations of the variation of the impact according to the frequency of the events involved. There are many situations where the influence of human activity will be less in relative terms for high magnitude events of low frequency. In the case of urban development this can be accounted for by the contrasts in infiltration properties between an urban and a natural surface being very much reduced under conditions of extreme wetness. A situation of this type is evident in Figure 5.6D which depicts hypothetical flood frequency curves for a small basin under natural and urban land use in the Palo Alto region of California derived by Crippen and Waananen (1969). In this case it is tentatively suggested that the curves converge at about the twenty-five-year recurrence interval. Hollis (1974), working in the Canons Brook catchment in the UK, similarly concluded that the effects of urbanization on floods with a return period of around twenty years, was minimal. Although this frequency threshold will vary from region to region in response to physiographic conditions it must be borne in mind as a highly significant qualification in any attempt to quantify man's impact.

5.5 QUALITY AS WELL AS QUANTITY

Traditionally, discussion of hydrological processes has focused on *quantity* considerations and assessment of the volumes of water passing through the various components of the hydrological cycle. More recently, however, interest has also been directed towards the physical and chemical *quality* of that water and the manner in which the quality reflects the hydrological processes involved and the catchment characteristics. This interest has been stimulated by the growing concern for environmental quality and pollution and the work of biologists in the field of nutrient cycling (e.g. Likens *et al.*, 1977). Against this background a number of catchment-based studies of man's effect on hydrological processes have included water quality parameters within their scope and this review can usefully include this additional facet of hydrological response.

When considering man-induced changes in water quality, a clear distinction must be made between modified quality and pollution, because many of the changes may be extremely significant from the hydrological and environmental viewpoint and yet not warrant being described as pollution. In turn, changes in water quality may not necessarily involve the addition of polluting substances. Furthermore, although man's impact on water quality necessarily involves the general problem of water pollution from domestic, municipal, and industrial effluents, discussion of such pollution must fall outside the scope of this account which is primarily concerned with hydrological processes. Discharge of effluent from a culvert or pipe, into a river channel, frequently termed *point-source pollution*, provides little opportunity for interaction with catchment-wide processes. Conversely, *nonpoint pollution* which has received increasing publicity in recent years and which includes the washing of fertilizers and pesticides from agricultural land into streams is necessarily reflected in the quality dimension of many of the hydrological processes operating in a drainage basin, and merits inclusion.

The drainage basin has frequently been selected as a convenient unit for evaluating the cycling of minerals and nutrients through terrestial ecosystems (e.g. Likens and Bormann, 1975). Associated measurements of the nutrient and mineral content of water from various locations within the drainage basin system have highlighted the complex interactions between precipitation input, vegetation uptake and release, biomass storage, rock weathering, and streamflow loadings, which are involved in regulating the quality dimensions of the hydrological cycle. Such ecosystems are delicately balanced and human intervention can cause important changes in water quality. Even man-induced changes in precipitation quality, related to atmospheric pollution and associated 'acid rain', can be significant. For example, Oden (1976) has shown that in areas such as Sweden with crystalline rocks of low buffering capacity, increased rainfall acidity has produced a clear negative trend in the pH and bicarbonate content of rivers and lakes , and a corresponding increase in sulphate concentrations.

5.5a Vegetation removal

More widespread are the changes in streamflow quality that may be induced by man's interference with the vegetation or soil cover of a catchment. Perhaps the most classic example of the potential impact of vegetation modification is the experiment carried out in the Hubbard Brook Experimental Forest in New Hampshire (Pierce *et al.*, 1970). Somewhat extreme in nature, this involved the cutting of all trees and woody vegetation on watershed 2 (Fig. 5.7A) during November and December 1965. This material was left in place and soil disturbance minimised. The following June an organic herbicide, bromacil, was sprayed over the basin by helicopter to kill all sprouting vegetation and manual spraying was repeated during the summers of 1967 and 1968 to prevent regeneration. The paired watershed approach was used to evaluate the effects on water quality, and in Figure 5.7A post-treatment streamflow quality is compared to that from catchment 6 with an undisturbed forest cover.

Because clearance occurred in winter, little effect was immediately apparent, but immediately after snowmelt commenced marked changes in streamflow chemistry occurred. Nitrate concentrations increased most and exhibited an overall increase of about fifty-fold during the three year period. In the control catchment the highest nitrate concentrations were observed in spring but this pattern was reversed on the treated basin. Cation concentrations also increased, although not to the same extent as nitrate (Fig. 5.7A), but sulphate levels were reduced. The great increase in nitrate levels can be

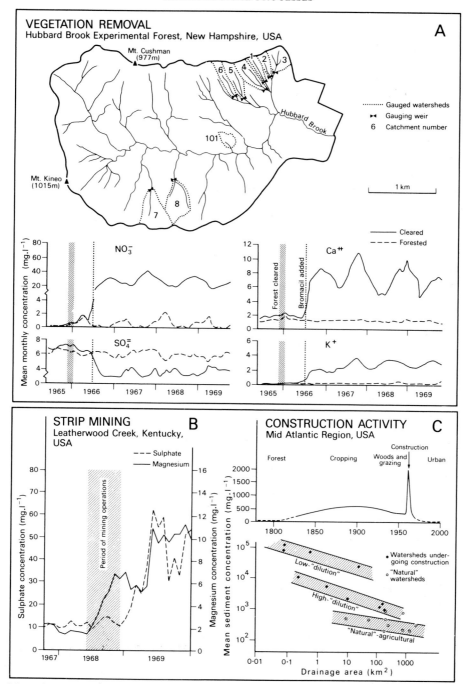

FIG. 5.7 MAN'S IMPACT ON WATER QUALITY
Results of the Hubbard Brook vegetation removal experiment (A) and of studies of the effects of strip mining (B) and construction activity (C) are illustrated. (Sources: (A) Pierce *et al.*, 1970; (B) Curtis, 1973; and (C) Wolman, 1967 and Wolman and Schick, 1967)

ascribed to the decomposition of organic matter and to the complete lack of nitrate uptake by vegetation. Being highly soluble, the nitrate ions were readily flushed to the stream. Increases in cation concentrations were closely linked to those of nitrate in that nitrification mobilises cations through the process of cation exchange. The reduced sulphate levels were attributed to the inhibition of sulphur-oxidising bacteria by the high nitrate concentrations, to the creation of anaerobic conditions within the soil by increased moisture content and to a general dilution associated with increased streamflow volumes. Overall, the total net load of dissolved inorganic material removed in the streamflow from the cleared watershed was about fifteen times greater than that from the control basin.

The treatment involved in this experiment was extreme in nature, but similar, if less marked, increases in nitrate concentrations have been documented in other conventionally clearcut catchments in New Hampshire, USA (Pierce *et al.*, 1972). For example, clearcutting watershed 101 at Hubbard Brook (Fig. 5.7A) caused an increase in nitrate levels about one-third of that described above. Nevertheless, care must be exercised in interpreting the results from New Hampshire as being generally applicable, because Aubertin and Patric (1972) report only small increases in nutrient levels after clearcutting in West Virginia. Similarly one can only speculate as to the impact of many other forms of vegetation change on streamflow quality.

5.5b Soil disturbance

Where human activity involves widespread disturbance of soil and regolith as well as vegetation removal, mineral and nutrient cycles will be further disrupted by acceleration of natural weathering and leaching processes. A useful example of this effect is provided by Figure 5.7B which illustrates some results of a study of strip mining on the streamflow quality of a small mountain basin in the Kentucky Appalachians, reported by Curtis (1973). Disturbance of 40 per cent of this seventy hectare catchment resulted in a nearly five-fold increase in sulphate and magnesium concentrations. In some mining areas the exposure of sulphide bearing minerals which weather to form ferrous sulphate and sulphuric acid can give rise to extremely acid streamflow and greater increases in solute levels, both as a result of the presence of the acid and its action in modifying weathering processes. Water quality problems associated with acid mine drainage are widespread in the eastern USA (Biesecker and George, 1966).

Surface disturbance will also generally produce increased erosion and therefore higher suspended sediment concentrations in streams. Although watershed sediment yields are often evaluated in terms of erosion rates and soil loss (e.g. Chapters 7 and 12), they also have important implications for water quality. Suspended sediment concentrations and associated turbidity levels exercise important controls over aquatic life and water use. Furthermore, the distinction between sediment and solute loads is somewhat arbitrary in view of chemical adsorption and exchange processes. A sizeable proportion of the output of inorganic phosphorus from a catchment can, for example, be transported adsorbed onto fine-grained sediment. The effects of various agricultural practices in increasing sediment loads are now well documented in the literature and Table 5.2 summarises some results collected by the United States Department of Agriculture which contrast annual soil loss under conditions of clean-tilled crops and dense vegetation cover for five locations.

Table 5.2

ANNUAL SOIL LOSS FROM FIVE CONTRASTING LOCATIONS IN THE USA UNDER CONDITIONS OF CLEAN TILLAGE AND
DENSE VEGETATION COVER

Location	Average annual precipitation (mm)	Annual soil loss (tonnes. ha^{-1})	
		Clean tillage	Dense cover
Bethany, Missouri	884	27·4	0·12
Tyler, Texas	1037	11·1	0·049
Guthrie, Oklahoma	838	9·7	0·012
Clarinda, Iowa	681	7·5	0·023
Statesville, N. Carolina	1149	9·0	0·005

Source: ASCE (1975)

Compared to dense vegetation cover, clean-tillage can cause increases in soil loss of as
much as 1800-fold,

5.5c Building activity

The greatest increases in sediment concentration that are documented in the literature
are, however, those associated with construction activity, and work undertaken in the
eastern USA (e.g. Guy, 1965; Wolman, 1967) provides a valuable perspective on this
aspect of human impact. Wolman and Schick (1967) report concentrations of 3000-
150000 mg.l^{-1} in streams draining construction sites in this area whilst in agricultural
watersheds the highest comparable concentration was 2000 mg.l^{-1}. Figure 5.7C compares
estimated discharge-weighted mean concentrations for natural watersheds and streams
draining construction sites in the region of metropolitan Baltimore and Washington, USA.
The inverse relationship between concentration and catchment area is related to the well-
documented decrease in sediment delivery ratio with increasing basin size. Whereas rural
catchments exhibit mean concentrations of approximately 500–600 mg.l^{-1}, the values for
basins where construction sites constitute a large proportion of their area, can be as high
as 25000 mg.l^{-1} or more. Figure 5.7C also attempts to interpret the classic diagram
produced by Wolman (1967) of the trend of sediment yields over the past 200 years in the
Piedmont region of the eastern USA in terms of the discharge-weighted mean concentration of a typical river draining that area. Prior to the development of agriculture,
concentrations are approximately 50 mg.l^{-1} and these increase to 600 mg.l^{-1} during the
period of intensive farming. With the decline in agriculture during the period immediately
preceding urban expansion, concentrations drop to 300 mg.l.$^{-1}$ but subsequently rise to
2000 mg.l^{-1} during the period of construction activity. Finally, with the establishment of
stable urban conditions, levels fall to less than 200 mg.l^{-1}.

5.5d Irrigation

Changes in hydrological processes associated with the development of irrigation can also
have important repercussions for water quality. Application of water for crop production
in semi-arid areas results in increased evapotranspiration and this can cause accumulation
of soluble salts within the soil. In addition, an increase in the level of the water table often

permits capillary rise of saline groundwater, again causing salt accumulation. Salinity problems of this type are one of the greatest difficulties facing the development of successful irrigation systems and can be overcome only by careful regulation of groundwater levels and by leaching of accumulated salts from the soil. These measures can have a considerable impact on rivers draining irrigated areas and receiving return water highly charged with soluble salts. Figure 5.8A illustrates this impact by comparing the existing quality characteristics during the irrigation season of the Lower Yakima River in Washington, western USA, with those that would probably exist in the absence of irrigation return flow. Sylvester and Seabloom (1963) estimate that sodium concentrations, which are particularly influenced by ion exchange mechanisms within the soil, show a fifty-fold increase.

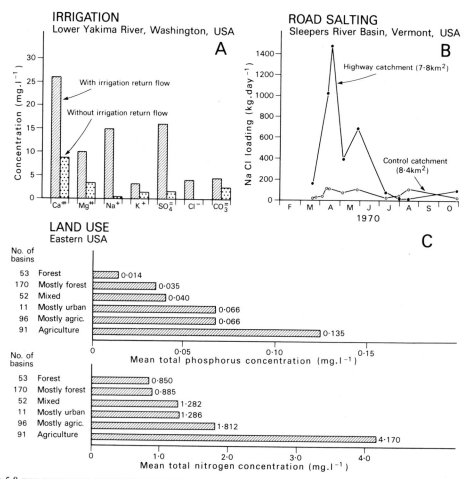

FIG.5.8 THE EFFECTS OF IRRIGATION DEVELOPMENT AND NONPOINT POLLUTION ON STREAMFLOW QUALITY
Results of three case studies discussed in the text are illustrated (Sources: (A) data in Sylvester and Seabloom, 1963; (B) Kunkle, 1972; and (C) Omernik, 1976)

5.5e Nonpoint pollution

To a certain extent, changes associated with irrigation return flow also reflect additions of fertilizer and other chemicals to irrigated areas, but the impact of additional inputs to catchment nutrient and mineral cycles is best seen in more obvious examples of nonpoint pollution from land use activities. In areas where severe winters necessitate the frequent and liberal use of road salt to maintain traffic flow, road drainage can constitute an extremely significant source of nonpoint pollution to adjacent streams. The total amount of de-icing chemicals used in the USA is probably close to 2 million tonnes per year and there is growing concern over the possible effects on receiving streams. Figure 5.8B, based on the work of Kunkle (1972) in Vermont, compares the daily NaCl loads of a catchment receiving highway drainage with those of an unaffected control watershed and illustrates the magnitude of the modification involved.

Perhaps the most significant form of nonpoint pollution and one that has attracted increasing attention in recent years, is the change in water quality caused by agricultural activity and more particularly the loss of nutrients to streams from applications of fertilizer and animal wastes. As a result of work carried out in a small watershed in Illinois, Kohl *et al.* (1971) suggested that as much as 50 per cent of the fertilizer applied could be lost to the streams. These findings have subsequently been questioned and other studies have documented instances where stream nutrient levels remain low in areas of very high fertilizer applications, but in spite of such controversy it is generally recognised that nonpoint agricultural sources provide a significant contribution to increased nutrient levels in streamflow. These may in turn cause excessive algal growth in river channels and where such streams feed lakes or reservoirs increased nitrogen and phosphorus loadings can give rise to problems of eutrophication.

Numerous small catchment studies aimed at evaluating the impact of agricultural sources on streamflow quality can be found in the literature, but specific reference can usefully be made to the more general findings of the National Eutrophication Survey established in the USA (e.g. Omernik, 1976). This has been investigating the relationships between stream nutrient loadings and watershed land use at the national scale. During 1972 and 1973 samples were collected from 473 small watersheds in the area of the USA east of the Mississippi River and some of the results obtained are presented in Figure 5.8C. Although embracing a variety of physiographic conditions, these results reflect a significant tendency for nutrient concentrations to be lower in streams draining forested watersheds than those draining areas used primarily for agriculture. Mean total nitrogen concentrations, for example, exhibit a five-fold difference between forested and agricultural watersheds and a large measure of this difference must derive from nonpoint agricultural sources and in turn reflect the impact of man.

5.6 PROBLEMS AND PROSPECTS

Studies of man's effects on hydrological processes can be readily justified academically in terms of the evidence they provide concerning his importance in the functioning of contemporary environmental systems. However, these effects are not without their problems and such studies can serve the further purpose of highlighting potential problems that may face man as his impact intensifies. These may include increased flood hazard and other changes in river regime, reduced availability of groundwater, deterioration of water quality and widespread eutrophication of water bodies and river systems in response to increased nutrient loadings. Figure 5.9 presents in simple diagrammatic form the example of a problem that has occured on Long Island, New York, as a result of effluent disposal practices and groundwater exploitation for water supply. Here groundwater outflow exceeded recharge and the negative balance caused widespread lowering of the water table and, more importantly, the incursion of saline water into the aquifers. Figure 5.9A depicts the early stages of water resource development on Long Island. Water supply wells were constructed in the shallow upper aquifer provided by the glacial deposits and wastewater was returned to the aquifer via cess pools and septic tanks. This situation maintained an approximate balance between outflow and recharge. However, the return of waste water to the shallow aquifer contaminated the water supply wells and these were eventually deepened to tap the lower aquifers (Fig. 5.9B). A negative groundwater balance was then initiated, since a proportion of the wastewater returned to the shallow aquifer discharged to streams and directly to the sea and did not penetrate the deep aquifer. Some reduction in recharge could also have been associated with the development of impervious conditions in more densely populated areas. Intrusion of saline water resulted. Subsequent development of an integrated sewerage system saw the replacement of septic tanks and cess pools by treatment works which discharged their effluent directly to the sea. Aquifer recharge was thereby drastically reduced and further saline intrusion resulted (Fig. 5.9C).

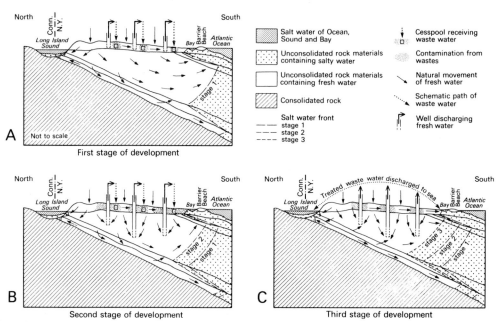

FIG.5.9 PROBLEMS CAUSED BY GROUNDWATER ABSTRACTION AND WASTEWATER DISPOSAL ON LONG ISLAND, NEW YORK
(Source: Cohen et al., 1968)

Because of the water supply problems associated with falling water tables and saline intrusions, a complex management programme has now been introduced to control and modify the functioning of the hydrological cycle on Long Island. This includes the use of recharge wells and diffusion basins to promote groundwater recharge and provides an excellent example of man's deliberate control of hydrological processes, a feature that will become increasingly common in the future.

Much empirical evidence concerning the impact of man on the hydrological cycle has been gathered over the past twenty years. It is to be hoped that this will be used by hydrologists in the future to provide guidelines for avoiding many of the problems that have resulted in the past and to develop models which will permit the long term prediction of the effects of human activity and the testing of alternative development and management strategies.

REFERENCES

ACKERMANN, W. C., 1969, 'Scientific hydrology in the United States', *The Progress of Hydrology, Proceedings of the First International Seminar for Hydrology Professors*, 1 (US National Science Foundation), pp. 50-60.

ADAMSON, C. M., 1974, 'Effects of soil conservation treatment on runoff and sediment loss from a catchment in southwestern New South Wales, Australia', *Effects of Man on the Interface of the Hydrological Cycle with the Physical Environment, Proceedings of the Paris Symposium*, IAHS Publ. 113, pp. 3-14.

ANDERSON, H. W., HOOVER, M. D. and REINHART, K. G., 1976, 'Forests and Water.' *US Forest Service General Technical Rept.*, PSW-18/1976.

ASCE, 1975, *Sedimentation Engineering* (American Society of Civil Engineers, New York).

AUBERTIN, G. M. and PATRIC, J. H., 1972, 'Quality water from clearcut land?' *Northern Logger*, 20, pp. 14–15, 22–3.

BATES, C. G. and HENRY, A. J., 1928, 'Forest and stream-flow experiment at Wagon Wheel Gap, Colo. Final report on completion of the second phase of the experiment.' *Monthly Weather Review*, Supplement No. 30.

BIESECKER, J. E. and GEORGE, J. R., 1966, 'Stream acidity in Appalachia as related to coal mine drainage.' *US Geol. Surv. Circular*, 526.

COHEN, P., FRANKE, O. L. and FOXWORTHY, B. L., 1968, 'An atlas of Long Island's Water Resources.' *New York Water Res. Committee Bull.*, 62.

COLLIER, C. R., PICKERING, R. J. and MUSSER, J. J. (eds.), 1970, 'Influences of strip mining on the hydrological environment of parts of Beaver Creek Basin, Kentucky, 1955–66.' *US Geol. Surv. Prof. Paper*, 427-C.

CRIPPEN, J. R. and WAANANEN, A. O., 1969, 'Hydrologic effects of suburban development near Palo Alto, California.' *US Geol. Surv. Open File Rept.*

CURTIS, W. R., 1973, 'Effects of strip mining on the hydrology of small mountain watersheds in Appalachia', *Ecology and Reclamation of Devastated Land*, ed. R. J. HUTNIK and G. DAVIS (Gordon and Breach, New York), pp. 145–57.

ENGLER, A., 1919, 'Einfluss des Waldes auf den stand der Gewasser.' *Mitt. Schweiz anst für das Forsliche Versuchswesen*, 12.

DILS, R. E., 1957, 'A guide to the Coweeta Hydrologic Laboratory.' *US Dept. of Agric., Forest Service Rept.*

GLEASON, G. B., 1952, 'Consumptive use of water, municipal and industrial areas.' *Trans. Amer. Soc. Agric. Engineers*, 117, pp. 1004–9.

GOLF, I. W., 1967, 'Contribution concerning flow rates of rivers transporting drain waters of open-cast mines.' *IAHS Publ.*, 76, pp. 306–16.

GUY, H. P., 1965, 'Residential construction and sedimentation at Kensington, Md.' *Proc. Federal Inter-Agency Sedimentation Conference, USDA, Misc. Publ.*, 970, pp. 30–7.

HARROLD, L. L., 1971, 'Effects of vegetation on storm hydrographs', *Biological Effects in the Hydrological Cycle. Proceedings of the Third International Seminar for Hydrology Professors* (US National Science Foundation), pp. 332–46.

HELVEY, J. D., 1972, 'First-year effects of wildfire from mountain watersheds in north-central Washington', *Watersheds in Transition, American Water Resources Association Proceedings Series*, 14, pp. 308–12.

HIBBERT, A. R., 1967, 'Forest treatment effects on water yield,' *Proc. Int. Symp. on Forest Hydrology*, ed. W. E. SOPPER and H. W. LULL (Pergamon, Oxford), pp. 527–43.

HILLS, R. C., 1969, 'Effects of agricultural treatment on the quantity of water in the soil', *The Role of Water in Agriculture, Aberystwyth Symposia in Agricultural Meteorology Memorandum*, 12, pp. F1–7.

HOLLIS, G. E., 1974, 'Urbanization and floods', *Fluvial Processes in Instrumented Watersheds, Institute of British Geographers Special Publication, No. 6*, ed. K. J. GREGORY and D. E. WALLING, pp. 123–39.

HOLSTENER–JORGENSEN, H., 1967, 'Influence of forest management and drainage on ground-water fluctuations', *Proc. Int. Symp. on Forest Hydrology*, ed. W. E. SOPPER and H. W. LULL (Pergamon, Oxford), pp. 325–34.

HOLTAN, H. N. and KIRKPATRICK, M. H., 1950, 'Rainfall infiltration and hydraulics of flow in run-off computation.' *Trans. Amer. Geophys. Union*, 31, pp. 771–9.

HOWE, G. M., SLAYMAKER, H. O. and HARDING, D. M., 1967, 'Some aspects of the flood hydrology of the upper catchments of the Severn and Wye.' *Trans. Inst. Brit. Geog*, 41, pp. 33–58.

KOHL, D. H., SHEARER, G. B. and COMMONER, B., 1971, 'Fertilizer nitrogen: contribution to nitrate in surface water in a Corn Belt watershed.' *Science*, 174, 1331–4.

KUNKLE, S. H., 1972, 'Effects of road salt on a Vermont stream.' *Jnl. Amer. Waterworks Assoc.*, 64, pp. 290–4.

LEOPOLD, L. B., 1968, 'Hydrology for urban land planning—a guidebook on the hydrologic effects of urban land use.' *US Geol. Surv. Circular*, 554.

LIKENS, G. E. and BORMANN, F. H., 1975, 'An experimental approach in New England landscapes.' *Ecological Studies*, 10, pp. 7–29.

LIKENS, G. E., BORMANN, F. H., PIERCE, R. S., EATON, J. S. and JOHNSON, N. M., 1977, *Biogeochemistry of a Forested Ecosystem* (Springer-Verlag, New York).

LVOVICH, M. I. and CHERNOGAEVA, G. M., 1977, 'The water balance of Moscow', *Effects of Urbanization and Industrialization on the Hydrological Regime and on Water Quality, IAHS Publ.*, 123, pp. 48–51.

McCUBBIN, G. A., 1938, 'Agricultural drainage in south-western Ontario, its effects on stream discharge.' *Engineering Journal*, 21, pp. 66–70.

MUSTONEN, S. E. and SEUNA, P., 1971, 'Metsäojituksen vaikutuksesta suon hydrologiaan,' *Vesientutkimuslaitoksen Julkaisuja*, 2.

NEW ZEALAND MINISTRY of WORKS, 1968, *Annual Hydrological Research Report for Moutere*, 1 (Wellington).
1971, 'Moutere, IHD experimental basin No. 8.' *Hydrological Research Annual Report*, 20.
1978, 'Research and survey, annual review 1977.' *Water and Soil Technical Publication*, 9.

NOBLE, E. L., 1965, 'Sediment reduction through watershed rehabilitation', *Proc. Federal Inter-Agency Sedimentation Conference. USDA Misc. Publ.*, 970, pp. 114–23.

ODEN, S., 1976, 'The acidity problem—an outline of concepts.' *Proc. First Int. Symp. on Acid Precipitation, US Forest Service Rept.*, NEFES/77–1, pp. 1–36.

OMERNIK, J. M., 1976, 'The influence of land use on stream nutrient levels.' *US Environmental Protection Agency, Ecological Research Series Rept.* EPA–600/3–76–014.

PEREIRA, H. C., 1973, *Land Use and Water Resources* (CUP, Cambridge).

PIERCE, R. S., HORNBECK, J. W., LIKENS, G. E. and BORMANN, F. H., 1970 'Effects of vegetation elimination on stream water quantity and quality', *Results of Research on Representative and Experimental Basins, Proceedings of the Wellington Symposium, IAHS Publ.*, 96, pp. 311–28.

PIERCE, R. S., MARTIN, C. W., REEVES, C. C., LIKENS, G. E. and BORMANN, F. H., 1972, 'Nutrient loss from clearcuttings in New Hampshire', *Watersheds in Transition, American Water Resources Association Proceedings Series,* 14, pp. 285–95.

RICHARDS, K. S. and WOOD, R., 1977, 'Urbanization, water redistribution and their effect on channel processes', *River Channel Changes,* ed. K. J. GREGORY (Wiley, London). pp. 369–88.

RODDA, J. C., 1976, 'Basin studies', *Facets of Hydrology,* ed. J. C. RODDA (Wiley, London), pp. 257–97.

RYCROFT, D. W. and MASSEY, W., 1975, 'The effect of field drainage on river flow.' *Min. Agric., Fisheries and Food, Field Drainage Experimental Unit Tech. Bull.,* 75/9.

SCARF, F., 1970, 'Hydrologic effects of cultural changes at Moutere experimental basin.' *Jnl. of Hydrology (N.Z.),* 9, pp. 142–62.

SHEVCHENKO, M. A., 1962, 'Effects of various methods of tillage on the reduction of snowmelt runoff from sloping land.' *Soviet Hydrology,* pp. 27–33.

SNYDER, C. T., FRICKEL, D. G., HADLEY, R. F. and MILLER, R. F., 1976, 'Effects of off-road vehicle use on the hydrology and landscape of arid environments in central and southern California.' *US Geol. Surv. Rept.,*WRI–76–99.

SYLVESTER, R. O. and SEABLOOM, R. W., 1963, 'Quality and significance of irrigation return flow.' *Proc. Amer. Soc. Civil Eng. Jnl. Irrig. and Drainage Div.,* 89, pp. 1–27.

WAANANEN, A. O., 1969, 'Urban effects on water yield', *Effects of Watershed Changes on Streamflow,* ed. W. L. MOORE and C. W. MORGAN (University of Texas), pp. 169–82.

WALLING, D. E. and GREGORY, K. J., 1970, 'The measurement of the effects of building construction on drainage basin dynamics.' *Jnl. Hyd.,* 11, 129–44.

WOLMAN, M. G., 1967, 'A cycle of sedimentation and erosion in urban river channels.' *Geografiska Annaler,* 49 Ser.A, pp. 385–95.

WOLMAN, M. G. and SCHICK, A. P., 1967, 'Effects of construction on fluvial sediment, urban and suburban areas of Maryland.' *Water Resources Res,* 3, 451–64.

6

Coastal Processes

E. C. F. BIRD*
University of Melbourne

6.1 INTRODUCTION

Coastal features are the outcome of a variety of processes working on the available geological materials in the zone where the land meets the sea. Cliffs and rocky shores have been shaped largely by erosional processes, and beaches, spits, and marshlands by depositional processes. The outlines of erosional and depositional features are related to the patterns of waves generated by winds blowing over the sea surface, and currents, particularly those associated with the rise and fall of the tide.

Geological materials available at the coast include both solid rock formations that outcrop in cliffs and along the shore, and unconsolidated sediments, ranging in size from large boulders and cobbles down through pebbles and sand to silt and clay. Some of the sediments have come from the breaking-down of coastal rock outcrops; others have been brought by rivers from the hinterland; and others have been washed in from the sea floor. Sand and pebbles are moved to and fro along the shore by waves and currents, and deposited as beaches and spits, whilst the finer sediments accumulate in sheltered inlets and estuaries as mudflats, which may be colonised by vegetation and built up as marshes. There is much variation in the way in which coastal processes operate, in the types of rock outcrop and unconsolidated sediment present, and in the kinds of landforms developing around the world's coastline. Accounts of the various relationships between landforms and processes can be found in textbooks of coastal geomorphology (e.g. Davies, 1972; King, 1972; Bird, 1976a; Komar, 1976).

Many coastal features have been shaped entirely by natural processes, but others have been modified, directly or indirectly, by man's activities. Some modifications take the

*Dr Bird is Chairman of the International Geographical Union's Commission on the Coastal Environment, and this paper uses examples taken from the Commission's project on world-wide shoreline changes (Bird 1976b).

form of either accelerating or slowing down of the natural course of coastal evolution, and these can be difficult to detect. Others, often more drastic, lead to features that would not have developed in the absence of man's activities. This essay examines some of the ways in which coastal processes have been modified by the impact of man.

6.2 DIRECT AND INDIRECT EFFECTS

There are not many examples of man's attempts to modify directly the processes at work in coastal waters. It is possible to disrupt wave motion, and weaken the energy of waves moving in towards a coast, by injecting 'air bubble curtains' that rise from perforated piping laid across the sea floor and fed with compressed air. This technique was used on a small scale to reduce wave action while engineering works were in progress at the entrance to Dover harbour several years ago, but it is too complicated and expensive to use as a means of reducing or halting coastal erosion.

More common are attempts to deflect or resist the effects of waves or currents by introducing structures such as sea walls, breakwaters, or groynes; by importing sediments to create or replenish beaches; or by planting vegetation. Some schemes have tried to halt coastal erosion; others have sought also to promote deposition. In each case, success depends on an understanding of the processes at work; the sources, quantities, and patterns of flow of coastal sediment; and the consequences of attempting to modify the coastal system in a particular way. Unfortunately, there have been failures. Around the world's coastline are many examples of derelict and abandoned structures, and elaborate works that remain unsuccessful. Some of these failures stem from assumptions that have been widely held by engineers and laymen, and are now considered questionable. These include the view that coastal erosion is always an evil to be resisted; that the natural condition of a coastline is one of stability, if not active accretion; and that, given the right design of structural works, accretion will automatically be induced on previously eroding shorelines. But coastal erosion in one place is commonly the source of coastal accretion in another, and the demand for anti-erosion works usually arises where there has been unwise development in the coastal zone. Evidence from a recent world-wide survey indicates that during the past century erosion by natural processes has been dominant on the world's beach-fringed coastlines (Bird, 1976b). And if sediment is not available in the coastal system, structural works cannot produce accretion.

Indirect effects of man's impact include those resulting from structural works elsewhere, usually on an adjacent part of the coast. Such works can lead to changes in nearshore topography, wave and current regimes, and patterns of coastal sediment flow. In addition, there have been coastal changes that can be traced to man's activities in the hinterland area, or offshore; changes which are the indirect outcome of activities that were not expected to modify coastal processes or alter coastline features. Some examples of each of these situations are given below.

6.3 EFFECTS OF SEA WALLS

The most obvious response to coastal erosion is to build a protective structure, usually a wall of masonry or reinforced concrete, designed to prevent waves from attacking the land margin. This is usually only justified where development has raised land values in the immediate hinterland, so that roads and buildings are in danger of being undermined if cliff recession continues. Sectors of cliffed coast at seaside resorts have generally been stabilized by the building of sea walls, usually incorporating some kind of promenade or public seafront area for recreational purposes.

Bournemouth, on the south coast of England, is a typical example. The seaside resort developed during the past century, spreading out from an initial 'watering place' at the mouth of the Bourne valley, and expanding laterally along the crests of receding cliffs of gravel-capped Tertiary sands and clays which lined the shore of Bournemouth Bay. These cliffs, subject to undercutting by storm waves and recurrent slumping, were fronted by a broad beach of sand and shingle derived from their erosion. A short section of sea wall, built in the form of a promenade in the nineteenth century, has gradually been extended east and west from the mouth of the Bourne valley, along the base of the cliffs, which were then graded into artificial coastal slopes and stabilized by planting vegetation (Fig. 6.1). There is now little left of the cliffs that once lined Bournemouth Bay, and the developed coastal property is no longer threatened by cliff recession.

The promenade at Bournemouth is appropriate for a seaside resort, and provides easy access to the beach, but as the sea wall extended the beach became depleted. The sand and shingle have been carried away, mainly eastwards, to Hengistbury Head and beyond, by the longshore drifting generated by south-westerly waves moving in from the English Channel, and without continuing natural replenishment from the eroding cliffs, the volume of beach material has dwindled. Insertion of a series of groynes extending out at right

FIG. 6.1 COASTAL ENGINEERING WORKS NEAR ALUM CHINE, BOURNEMOUTH
(E. C. F. Bird)

angles to the sea wall, breaking the beach into distinct compartments, reduced the rate of loss of sand and shingle, but the wastage continued. Here, as elsewhere, it was found that erosion became severe when waves were able to break against the sea wall during storms. Such waves were reflected, and generated a scouring of beach material offshore beyond the limits of the groynes. While some of this beach material was later carried back to the shore by gentler wave action in calmer weather, a proportion moved away eastwards as the result of the unhindered longshore drifting out beyond the ends of the groynes. By the 1960s the Bournemouth beaches had been depleted to such an extent that the local authority decided to introduce artificial beach nourishment, using sand and gravel dredged from the sea floor, and piped in to the shore (Fig. 6.2).

Depletion of beaches as a sequel to sea wall construction along the base of eroding cliffs can be illustrated at other seaside resorts in Britain and elsewhere. In Australia, for example, the cliffs of Tertiary sandstone and clay which used to line the north-eastern shores of Port Phillip Bay, where the coastal suburbs of Melbourne have developed, have also been largely stabilized by sea wall construction and graded to artificial slopes; and here, too, the beaches have been depleted. The effects of wave reflection by sea walls, also evident here, are a good illustration of the way in which an introduced structure can modify the processes at work on a coast.

FIG. 6.2 BEACH NOURISHMENT IN PROGRESS AT BOURNEMOUTH
(E. C. F. Bird)

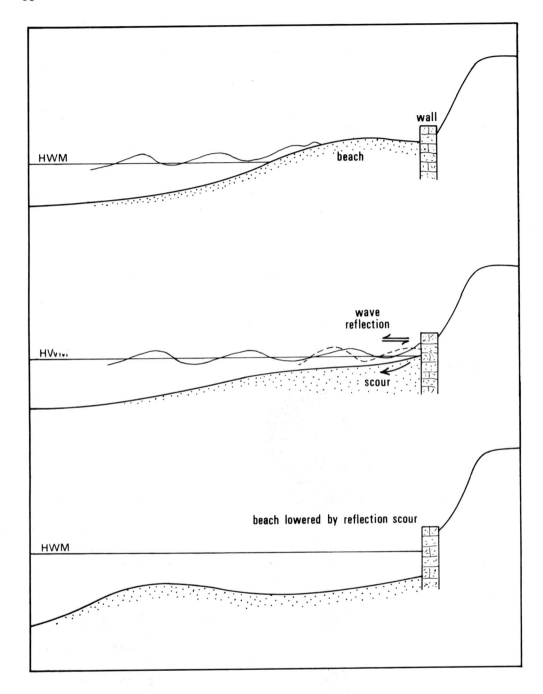

FIG. 6.3 SEA WALLS AND EROSION
A broad, high beach prevents storm waves breaking against a sea wall and will persist, or erode only slowly, but where ʼaves are reflected by the wall, scour is accelerated, and the beach is quickly removed.

Reflection scour begins only when storm waves break directly against a sea wall. As long as a broad protective beach is maintained in front of the sea wall the process is not effective, but if the sediment supply is cut off, and the beach becomes depleted so that storm waves can reach the sea wall, then the process of reflection scour sets in, and beach erosion accelerates. (Fig. 6.3).

Coastal engineers have experimented with a variety of designs and materials in sea wall construction in the hope of producing structures which do not set up reflection scour: for example, permeable groynes, and heaps of boulders or caged stones (gabions) placed along the eroding shoreline, intended to absorb storm wave energy rather than reflect it. But a simple structural solution to the problem of halting erosion of the land margin without losing the adjacent beach has proved elusive.

On some coasts the construction of sea walls has followed the domino principle: the halting of cliff erosion on one sector has been followed by beach depletion, allowing larger waves to move in and intensify erosion on the adjacent, unprotected sector; and when the sea wall has been extended to combat this, cliff erosion is accentuated on the next adjacent sector. The sequence of sea wall construction at Point Lonsdale, at the entrance to Port Phillip Bay, is an example (Fig. 6.4). In 1900, a sea wall was built to stabilize the shore in front of the township of Point Lonsdale, at the southern end of a gently curving sandy bay. The beach became depleted, and as each adjacent sector has shown intensified erosion the wall has been extended northwards, producing a sequence of slightly offset structures which commemorate the engineering ideas of the past seventy years, culminating in the recent addition of a large boulder wall (Fig. 6.5). If this approach continues, the whole of Point Lonsdale Bight will eventually consist of a walled shore, fronted by a depleted beach. On the other hand, if there had been no development immediately behind the Point Lonsdale coast, sea wall construction would have been unnecessary, and the relatively slow processes of natural erosion would have maintained a broad beach along this bay shoreline, a recreational and scenic asset conserved at the cost of secular losses of the sandy hinterland. Such losses would in any case have been offset by gains in sandy depositional land at Queenscliff, to the north. It is up to coastal zone managers to ensure that development is not permitted behind eroding sandy shorelines; for the existence of such development leads to the demand for erosion control, the addition of artificial structures, and the consequent depletion of the beach.

Where sea walls (dykes) have been built to enclose tidal marshlands and mudflats as a prelude to land reclamation, as in the Netherlands and on the shores of the German and Danish Wadden Sea, they are fronted by broad, gently-shelving intertidal areas which reduce the effects of storm waves, so that reflection scour does not usually take place. Instead, there is often continued accretion of mud, colonised by marsh plants, on the seaward side of such walls, and in places beaches of shelly sand accumulate at about high tide level. By excluding tidal waters, such sea walls modify the range and extent of tidal rise and fall, and may change the pattern and strength of current flow associated with tidal movements. At the mouth of the Rhine, tidal channels have locally been deepened as the result of changes in configuration due to the enclosure of formerly tidal areas by walls and dykes, and in some places the deflection of channel flow has led to scour on shorelines where it had not previously occurred.

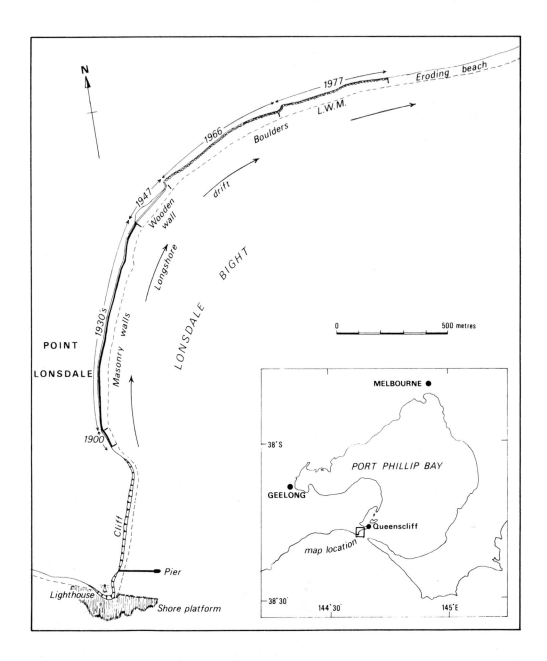

FIG. 6.4 SEQUENCE OF SEA WALL CONSTRUCTION AT POINT LONSDALE, VICTORIA, AUSTRALIA
(See Fig. 6.5)

6.4 EFFECTS OF BREAKWATERS

Breakwaters built out from the coast to provide a harbour sheltered from the effects of strong wave action may intercept the longshore drift of beach material that occurs where waves arrive at an angle to the shoreline. On the south coast of England, for example, there is eastward movement of sand and shingle as the dominant south-westerly waves move in to beaches of southerly aspect, and on the east coast of England the drift is southward in response to the dominant north-easterly waves arriving from the North Sea. Migration of beach material tends to deflect the mouths of rivers: the Adur and the Ouse in Sussex have had a long history of eastward deflection, while the East Anglian rivers have been diverted southwards by the growth of such spits as Orford Ness at the mouth of the Ore. Such deflections impede navigation, and where the river mouths are used as harbours it is necessary to stabilize a navigable entrance by means of breakwaters.

The harbour at Newhaven, on the mouth of the Sussex Ouse, has been maintained since 1731 by building and later extension of a breakwater on its western flank, to exclude south-westerly waves from the river mouth, and prevent the drifting of sand and shingle into the harbour entrance. As a result, beach material has accumulated to the west of the breakwater. Such accretion on the updrift side of a structure is generally followed by the onset of a sediment deficit, marked by beach depletion downdrift. East of the stabilized river mouth at Newhaven the beach at Seaford has been reduced, and cliff erosion on

FIG. 6.5 BOULDER WALLS NEAR POINT LONSDALE, VICTORIA
(J. McArthur)

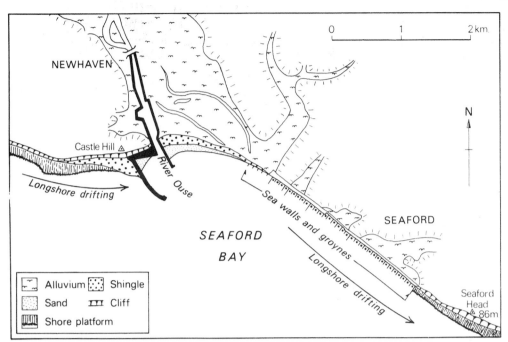

FIG. 6.6 ACCRETION OF DRIFTING BEACH MATERIAL ON THE WESTERN FLANK OF THE HARBOUR BREAKWATER AT
NEWHAVEN, SUSSEX, AND INTENSIFIED EROSION ALONG THE DRIFT-STARVED SECTOR TO THE EAST AT SEAFORD

Seaford Head accelerated. The coastline at Seaford has been maintained only by the
building of sea walls of masonry and concrete, to form an esplanade which is subject to
recurrent storm damage as the beach dwindles (Fig. 6.6).

This pattern of updrift accretion and downdrift erosion as a sequel to breakwater
construction is seen at many such harbours around the world. On the Florida coast, there
has been accumulation of southward-drifting sand alongside the breakwaters built to
stabilize South Lake Worth Inlet and give navigable access to a lagoon behind the coastal
sand barrier. On the Nigerian coast, the breakwaters built to stabilize the entrance to the
harbour in Lagos Lagoon have accumulated eastward-drifting sand on Lighthouse Beach
to the west and induced rapid erosion on Victoria Beach to the east. At Durban, in South
Africa, and Madras, in India, northward drift of sand has been arrested to prograde the
shoreline south of the harbour breakwaters, with beach erosion ensuing to the north; and
at Santa Barbara, in California, a breakwater has intercepted sand drifting from the
north and depleted the beaches to the south. On the east coast of Australia the
breakwaters built to stabilize the mouth of Tweed River have intercepted sand drifting
northward and resulted in beach depletion at the seaside resort of Coolangatta, to the
north. In some cases an attempt has been made to solve the problem by introducing sand
by-passing schemes, whereby some of the sand accumulating on the updrift side is pumped
through pipes under the harbour entrance and used to replenish the eroding beach
downdrift: such schemes are active at Durban; at South Lake Worth Inlet; at Surfside,
near Los Angeles; and at Salina Cruz, in Mexico.

A somewhat different pattern has developed at Lakes Entrance, in south-eastern Australia, where breakwaters were built alongside an artificial entrance cut through the coastal sand barrier to provide a permanent navigable channel into the Gippsland Lakes, replacing an earlier, impermanent natural outlet which was frequently deflected, and occasionally closed, by sand accretion (Bird, 1978). After the artificial entrance opened in 1889, sand accretion began on either side of the twin breakwaters, and led to the growth of cuspate forelands linked by an underwater sand bar which developed off the new cut. In this case, there have been alternations of eastward drifting resulting from south-westerly wave action, almost balanced by westward drifting produced by south-easterly wave action, and natural by-passing of sand probably takes place in either direction along the underwater sand bar. The breakwaters have thus modified coastal processes by establishing a localized nearshore outflow from the Gippsland Lakes which has interacted with wave action and available sediment in such a way as to develop a sand bar off the entrance. This bar, in turn, has established a new pattern of wave refraction, which has shaped the cuspate forelands east and west of the entrance (Fig. 6.7).

On the eastern shores of Port Phillip Bay (Fig. 6.4), seasonally alternating patterns of beach drifting have posed problems in harbour maintenance. The embayment at Hampton, for example, is subject to northward drifting of beach material in summer and southward drifting in winter, in response to the predominance of westerly wave action in the summer months, and south-westerly in the winter season (Fig. 6.8). Offshore breakwaters constructed in 1909, and extended in 1935–9, caused some sand accretion on the adjacent

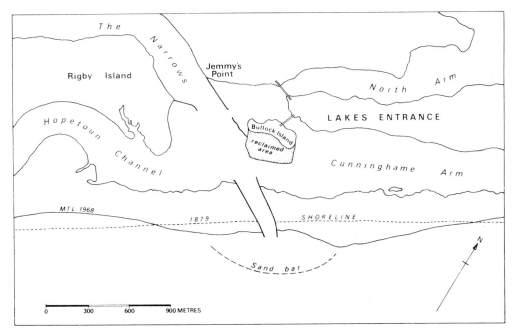

FIG. 6.7 THE SHORELINE AT LAKES ENTRANCE, VICTORIA, AUSTRALIA IN 1968
This shows the pattern of accretion on either side of the artificial entrance. The 1879 shoreline, mapped ten years before the artificial entrance was opened, is a basis for measuring subsequent changes

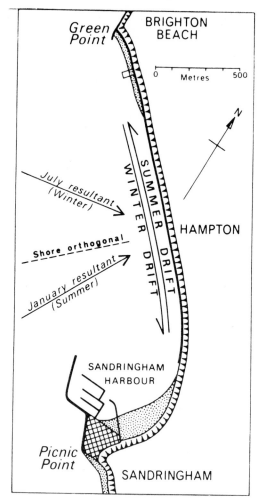

FIG. 6.8 COASTAL COMPARTMENT AT HAMPTON, PORT PHILLIP BAY, VICTORIA
Onshore wind resultants, and the seasonal pattern of beach drifting produced by consequent wave action are illustrated. Following the building of Sandringham Harbour (Fig. 6.9) the beach that formerly lined this coastal compartment has been dispersed by seasonal drifting, the bulk of the beach material having been deposited within the Harbour

shore, but when a large stone breakwater was built (1949–54) to shelter Sandringham Harbour, the sand moving southward each winter was trapped in its lee, within the sector no longer exposed to the south-westerly waves that previously generated northward drifting. As a result, the harbour has become partly filled with sand (Fig. 6.9).

An offshore breakwater has also been used at Santa Monica, in California, to protect a harbour adjacent to the pier. Built in 1934, this has modified the patterns of waves approaching the coast in such a way as to converge them to the lee and shape the deposition of a sandy cuspate foreland (Fig. 6.10). In a similar way, cuspate forelands have developed in the lee of ships grounded offshore, for example in Sukhumi Bay on the

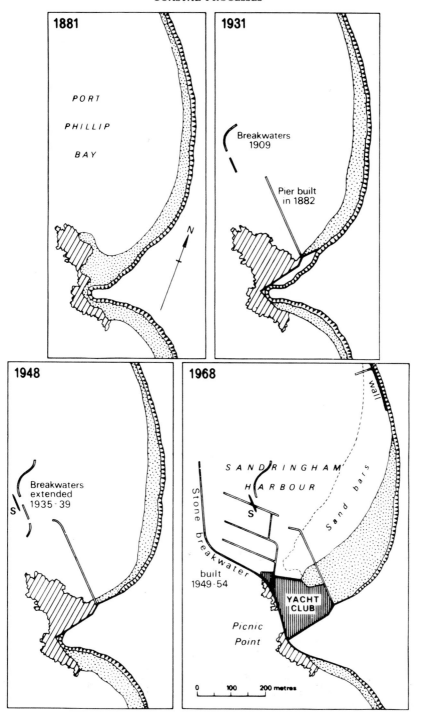

FIG. 6.9 STAGES IN THE EVOLUTION OF SANDRINGHAM HARBOUR AND ASSOCIATED SHORELINE CHANGES
(See Fig. 6.8)

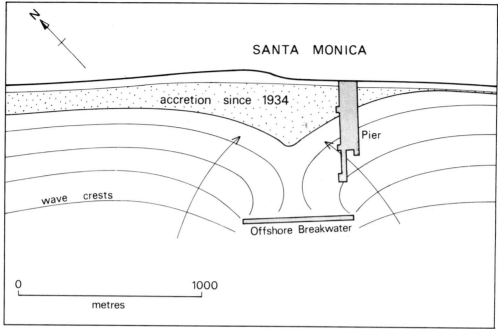

FIG. 6.10 SHORELINE ACCRETION AT SANTA MONICA, CALIFORNIA, FOLLOWING THE INSERTION OF AN OFFSHORE BREAKWATER IN 1934

Soviet Black Sea coast (Zenkovich, 1967: Fig. 193), and at Port Hueneme in California. In each case, localized trapping of sand has been followed by beach erosion downdrift, and at Santa Monica it has been necessary to replenish artifically the depleted beaches to the south on the Los Angeles coast.

Breakwaters built in the nineteenth century to enclose a bay area as a harbour, as at Portland on the south coast of England, can modify the wave regime and intensify wave refraction in such a way as to change the pattern of wave energy, and also the angle of wave incidence on the adjacent coast. This may explain the erosion of beaches and the redistribution of shingle on the shores of Weymouth Bay, to the north-east, in the period following the construction of the Portland breakwaters. Similar changes have followed the construction of an enclosing breakwater to form a harbour at another Portland, in south-eastern Australia, the sequel here being the onset of severe beach erosion at Dutton Way, to the north, where the energy of refracted waves reaching the shore has apparently increased (Fig. 6.11).

6.5   EFFECTS OF DREDGING AND DUMPING

In many places the approaches to ports have been dredged to provide or improve a navigable channel. Dredging can modify the pattern and velocity of currents, which in these environments are generated mainly by tidal action and river discharge. As a rule, deepening of a channel is likely to be followed by a weakening of the currents, and where

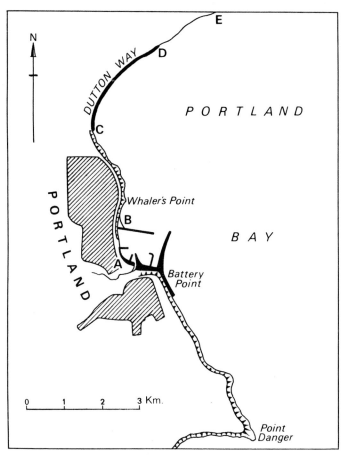

FIG. 6.11 COASTAL FEATURES AT PORTLAND, VICTORIA, AUSTRALIA
The construction of the large harbour breakwater at Battery Point in 1957–59 was followed by rapid erosion of the shoreline at Dutton Way (CD). This is now lined by a boulder wall, but shoreline erosion has intensified in the next sector (DE)

the approach is to an estuarine or lagoonal system the increased cross-sectional area of the dredged channel may also modify the range and extent of tidal rise and fall within that system. On the other hand, a channel dredged on an alignment which is shorter than the course of pre-existing natural channels may intensify current flow, and increase tidal ventilation within an estuary or lagoonal system.

Material dredged from such channels is commonly dumped either offshore, or in adjacent shallows, or carried onshore for use in land reclamation or beach nourishment works. Dredged material dumped offshore creates a new sea floor topography, which can modify wave patterns and the direction and strength of current flow, whereas material dumped onshore establishes a new coastal outline. In either case, there is usually dispersal and reworking of dredged material. An example is the Cairns area, in north-eastern Australia, where a channel dredged through extensive mudflats gives access to a port at the mouth of Trinity Inlet. The dredged mud, dumped in adjacent shallows, has been dispersed across Cairns Bay by wave action, and some of it has been deposited on bordering sandy shores, as at Machan's Beach and at Yarrabah in Mission Bay.

Dredging of the sea floor to obtain sand and gravel or other mineral resources results in localized deepening of the water, which enables larger waves to move in towards the adjacent shoreline. This can result in a change from progradation or stability to erosion, or the acceleration of erosion that was already in progress.

Erosion has become severe on the western shores of Botany Bay, near Sydney, Australia, following the dredging of a deeper channel through the bay entrance, which permits larger waves to move through into the bay, and break upon the western shoreline. The situation has been complicated by the extension of the airport runway out into the sea, a protrusion that may have intensified wave energy on the adjacent sandy shoreline in much the same way as the Portland breakwaters.

6.6   EFFECTS OF SEDIMENTOLOGICAL CHANGES

Apart from modifications to coastal processes caused by the addition of artificial structures there have been changes resulting from man-induced alterations in the nature and quantity of sediment available at the coast. Some of these have been direct, as in the cases where waste from coastal quarries spills directly into the sea, and is reworked by waves and currents to build beaches that would not otherwise have formed. An example of this is the chalk quarry at Hoed, on the east coast of Jutland in Denmark, where during the past two centuries unwanted flint nodules separated from the chalk have been dumped on the coast. The outcome has been local progradation of a beach ridge plain (Fig. 6.12).

In recent years, depletion of beaches, particularly at seaside resorts, has prompted coastal engineers to attempt artificial replenishment of the lost sand and shingle with material brought from inland or offshore sources. Mention has been made of the restoration of eroded resort beaches downdrift from harbour breakwaters at Durban and Santa Monica. On the Soviet Black Sea coast, beaches were formerly quarried to supply sand and gravel for building purposes, but by the early 1960s it was obvious that this extraction of beach sediment was resulting in accelerated erosion. The coast at Sochi, for example, began to suffer severe storm wave erosion as its beach became depleted. The procedure was then reversed, with loads of sand and gravel from inland quarries being brought down to the coast and dumped on the shore to restore the beach, improving a recreational resource at the same time as countering storm wave erosion.

In the United States there have been a number of beach nourishment projects, based on the dredging of sand from the sea floor and its delivery by pumping or dumping on to the eroding shore. The beach at Atlantic City, New Jersey, was restored in this way between 1935 and 1943; those at Palm Beach, in Florida, and West Haven, Connecticut, in 1948; and others subsequently at Virginia Beach, south of Cape Henry; Harrison County, Mississippi, and several sites on the coast of California. Mention has already been made of the artificial replenishment of Bournemouth beach, in southern England, with sand brought in from the sea floor, and similar projects are active elsewhere, for example at Mentone, on the shores of Port Phillip Bay, Australia (Fig. 6.13).

Less direct have been the effects of deforestation and cultivation of hinterlands within

FIG. 6.12 PROGRADED BEACH RIDGES SOUTH OF GRAVEL WASTE TIPS AT HOED QUARRY, JUTLAND, DENMARK
(F. Hansen)

the catchments of rivers draining to the coast. As a rule, such rivers show increased discharge and become more flood-prone; they deliver larger quantities of sediment to the coast, often with an increased proportion of coarser material. The outcome is accelerated growth of deltas, and progradation of beaches supplied with fluvial sands or gravels. It is believed that the reduction of vegetation cover by clearing, burning and grazing in the Mediterranean region, which became extensive about 2000 years ago, resulted in rapid siltation of bays and inlets at and near river mouths, especially on the Adriatic coast between Ravenna and Trieste, and on the shores bordering the Gulf of Corinth and the Gulf of Euboea, in Greece. There is, however, some doubt about the often-quoted rapid historical advance of the Mesopotamian deltaic coastline, at the mouths of the Tigris and

FIG. 6.13 ARTIFICIAL BEACH EMPLACED IN FRONT OF THE SEA WALL AT MENTONE, PORT PHILLIP BAY BY THE
METHOD SHOWN IN FIG. 6.2
(N. Rosengren)

Euphrates rivers (cf. Davis, 1956). More recently the onset of severe soil erosion in river
catchments around Chesapeake Bay has been correlated with shallowing and marsh
encroachment within that bay, an effect which is repeated on a larger scale in the humid
tropics, especially at Djakarta Bay in Indonesia, and on some of the Pacific Islands. On
the west coast of New Caledonia, for instance, there has been rapid extension of mudflats
and mangrove swamps in recent decades within bays and inlets receiving augmented
supplies of silt and clay from rivers that drain recently deforested catchments.

The sequence has been reversed, however, where dams have been constructed to
impound reservoirs, which act as sediment traps (cf. Chapter 8). Diminished sediment

yield from rivers thus modified has resulted in a reduction of sand supply to the beaches of Southern California. These beaches show southward longshore drifting, the sand being lost into the heads of submarine canyons close inshore, as at La Jolla, and as the fluvial input is reduced the beaches are becoming depleted. Several of the world's major deltas now show rapid shoreline erosion, the sediment supply that formerly maintained and prograded them having diminished as the result of dam construction. The most notable is the Nile, where the delta shoreline is receding rapidly, up to 40 m a year, erosion having accelerated in consequence of the reduction in sediment supply following the completion of the Aswan High Dam in 1970.

Reduction of fluvial discharge by dam construction is thought to have been a primary cause of the lowering of level of the Caspian Sea, which fell by 2·67 m between 1930 and 1975. In consequence, there has been extensive emergence of bordering shores, accompanied by widespread deposition, as the sphere of coastal processes has been withdrawn from the earlier shoreline. This seems to be a major example of the indirect impact of man on a coastline. Similar changes are taking place in the Aral Sea.

6.7   EFFECTS OF VEGETATION

Coastal processes can also be modified as the result of changes in vegetation. On the one hand, man's activities can reduce the extent and vigour of the vegetation cover, and result in the onset of erosion of terrain that was previously stabilized. On the other, the introduction of species that were not previously present, or the improvement of vegetation cover by artificial means (such as the addition of fertilizers) can influence the pattern of sedimentation and promote stability.

The first sequence is well illustrated where coastal dunes, previously stabilized by a cover of grasses, scrub or woodland, become unstable as the result of a reduction in vegetation cover, for example by deliberate clearance, burning, grazing by introduced animals (sheep, cattle, goats, rabbits), trampling, or vehicle damage. In south-eastern Australia the coastal dune fringe is typically interrupted by blowouts where the vegetation cover has been removed or destroyed, and sand is being moved inland by onshore winds. Some of these may have developed naturally behind eroded beaches, but many have alignments that originated where deflation began along trackways to the shore where trampling had weakened the vegetation cover. In some sectors the vegetation has become so weakened that the dunes have developed into mobile sand sheets: the active dunes at Cronulla, south of Sydney, exemplify this.

The sequence is reversed in places where coastal dunes that were previously active (either naturally, or because of prior destruction of vegetation) have been stabilized by the planting of retentive vegetation, notably the European marram grass, *Ammophila arenaria*. The introduction of marram grass to the rear of broad sandy beaches has been followed by the accretion of sand blown from the beach to form high foredunes. Elsewhere, previously mobile dune topography has been arrested by planting grasses, shrubs, and forest trees, for example in the Landes region of south-western France, at Culbin on the Scottish coast, and in the Danish 'Klit', the coastal fringe of stabilized dune topography where a century ago the sand was spilling inland across farming country. Generally such stabilization of drifting coastal sands has been regarded as beneficial, but on the Oregon coast the development of high foredunes as a consequence of marram grass introduction has recently been criticized on the grounds that it has greatly diminished sand supply to the hinterland, so that the area of drifting sand dunes has been much reduced. In 1977 there were demands for the removal of marram grass and the remobilization of coastal sands to ensure the preservation of active dune areas as open spaces within the Oregon Dunes National Recreation Area.

Vegetation has a similar effect in coastal marshlands, tending to promote the accretion of sediment within colonized zones in such a way as to build relatively stable salt marshes. In 1870, hybridization between a European and an American species of *Spartina* in the marshlands of Southampton Water yielded a new and vigorous species, now known as *Spartina anglica*. This species was able to spread forward on to mudflats below the level of other salt marsh species, and its vigorous growth provided a filtering matrix which rapidly accreted sediment to build a new prograded salt marsh terrace. It was soon introduced to Poole Harbour and other British and European estuaries and inlets, and eventually to similar areas as far away as the Tamar in Tasmania, Anderson's Inlet in south-eastern Australia, and several New Zealand estuaries, to act as an agent in marshland development, preparing the way for reclamation. It provides a good example of the impact of man on coastal processes, for the dense growth of *Spartina* greatly diminishes wave and current action as the tide rises and falls in marsh-fringed estuaries, with the result that sedimentation is facilitated within the *Spartina* zone. In some estuaries, however, the *Spartina* terraces built during the past century are now showing decay and erosion, a consequence of marginal die-back of the *Spartina*, for reasons which have not been precisely determined. At Lymington, for example, the muddy sediment previously trapped by *Spartina* growth is now being laid bare and eroded away by wave and current scour along the channel margins.

6.8   CONCLUSIONS

The various examples quoted show how coastal processes can be modified by the addition of artificial structures, by changes in the nature and quantity of available sediment, and by modifications in the vegetation cover, all due directly or indirectly to man's activities. In some sectors the modifications have resulted in deposition and coastal progradation that would not otherwise have occurred; in others, they have initiated or accelerated erosion. Sequences have been noted where man's impact has stimulated coastal deposition and further interference has resulted in erosion of newly-deposited terrain; or where an initial erosional response has subsequently been reversed. Man's impact on coastal processes has been strongest and most obvious on densely-populated and intensively-used coastlines, but minor and indirect consequences of human interference may be detected elsewhere, particularly on coastal dune systems, which may show man-induced erosion (by way of stock grazing on dune vegetation, for instance) in areas remote from intensive usage. The presence of such erosional features in coastal dunes on sparsely-populated sectors of the Australian coastline, such as Discovery Bay, Cape Howe, and Fraser Island, shows how widespread are some of the impacts of man's activities on coastlines.

REFERENCES

BIRD, E. C. F., 1976a, *Coasts* (Australian National University Press, Canberra).

1976b, *Shoreline Changes During the Past Century: a Preliminary Review* (IGU Working Group on the Dynamics of Shoreline Erosion, Melbourne).

1978, *The Geomorphology of the Gippsland Lakes Region,* Ministry for Conservation, Victoria, Publication No. 186.

DAVIES, J. L., 1972, *Geographical Variation in Coastal Development* (Oliver and Boyd, Edinburgh).

DAVIS, J. H., 1956, 'Influences of man upon coast lines', *Man's Role in Changing the Face of the Earth*, ed. W. L. THOMAS (Chicago Univ. Press, Chicago), pp. 504–21.

KING, C. A. M., 1972, *Beaches and Coasts* (Arnold, London).

KOMAR, P., 1976, *Beach Processes and Sedimentation* (Prentice Hall, New Jersey).

ZENKOVICH, V. P., 1967, *Processes of Coastal Development*, ed. J. A. STEERS (Oliver and Boyd, Edinburgh).

Lithosphere

7

Slopes and Weathering

M. J. SELBY
University of Waikato

7.1 INTRODUCTION

Most of the land surface is composed of hillslopes and, except in areas of great aridity, low temperature and extreme steepness, they have a mantle of weathered rock and soil. By comparison with the underlying rock, soil is a weak and permeable material which can be readily modified by processes of erosion and by direct human interference. The processes of weathering which convert resistant rock to weaker soil are thus essential forerunners of erosion.

The results of weathering can be appreciated from a consideration of the characteristics of a deep profile (Table 7.1). The depth of weathering profiles and the completeness of the horizons are extremely variable. On ancient surfaces of stable continents profile depths may exceed 100 m in the humid tropics, but such depths are rare and 30 to 50 m is a more common maximum depth. Profile depths decrease as slope angles increase so that in many humid temperate environments slopes steeper than about 45° have either bare rock or very thin soil profiles. In some parts of the humid tropics, where weathering rates are very high, slopes with a vegetation cover may have angles of up to 70°, although they are then subject to periodic landsliding. Such high angles are relatively common in Tahiti and Papua New Guinea.

The description of weathering profile features given below (Table 7.1) assumes that the soil is developed *in situ*; where it is derived from transported materials the profile characteristics may be far more variable and the trend of strength and permeability with depth may be highly irregular.

Table 7.1
AN IDEALIZED WEATHERING PROFILE WITH SOME CHARACTERISTIC PROPERTIES

VI	Residual soil at the surface	No trace of the original rock is preserved; strength is lowest and permeability is greatest unless clay minerals have been formed.
V	Completely weathered	Original rock fabric may be preserved but colour and mineral composition are those of soil not rock.
IV	Highly weathered	More than 50 per cent of the material is soil. Corestones are common and rock fabrics and structure are dominant. Joints may be open and are then zones of weathering.
III	Moderately weathered	More than 50 per cent of the material is rock. Corestones are fitting; joints are partly open.
II	Slightly weathered	Discolouration is largely along joints.
I	Fresh rock	Joints are closed; rock is not discoloured.

Rock compressive strength, that is resistance to crushing, is measured in newtons per square metre and ranges from about 200 $MN.m^{-2}$ for a hard igneous rock to about 20 $kN.m^{-2}$ for a soft moist clay

Source: Dearman (1974)

On hard rocks the depth of weathering sets a limit below which human activity can modify the landscape with ease and with little expense. Soft rocks, such as unconsolidated sands or clays may have strengths no greater than those of soils and they also are easily modified by human activity – whether this is by deliberate intervention with machinery or from the, usually unwanted, results of induced erosion.

7.2   FACTORS CONTROLLING EROSION

The factors controlling the rate and type of erosion occurring on hillslopes include climate, topography, rock type, vegetation, and soil character (Fig. 7.1). They are linked in a web of relationships in which climate and geological factors are the most independent, with soil character and vegetation cover being dependent upon them, and closely related to each other. Vegetation, for example, is closely controlled by the amount, duration and intensity of rainfall; by the heat available during the growing season; and also by the capacity of soil to supply nutrients and an anchorage for plant roots. In its turn vegetation influences soil through the action of roots, the provision of organic matter, the uptake of nutrients and by providing strength which resists the impact of raindrops and the shear stresses imposed by running water and landsliding.

The extent to which human interference can modify the impact of the factors of erosion is varied. Rock type and topography cannot be altered on a large scale except at enormous cost, but on a minor scale the formation of agricultural terraces can reduce slope angles and shorten the effective length of slopes. The effect of climate can be most readily modified by reducing the impact of raindrops upon soil (See also Chapter 12).

A vegetation cover or a mulch of plant material, or plastic sheeting, can reduce splash erosion to negligible amounts. By far the easiest way to modify erosion is by manipulating plant cover. Thus man can greatly increase potential erosion by exposing soil, or reduce the erosion hazard to very low values by keeping a complete vegetation cover on the soil.

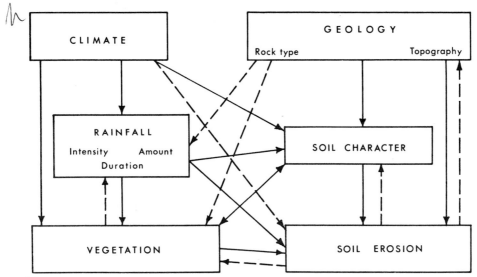

FIG. 7.1 RELATIONSHIPS AMONGST THE FACTORS INFLUENCING SOIL EROSION

Erosion is a function of the eroding power, that is the erosivity, of raindrops, running water and sliding or flowing soil masses, and the erodibility of the soil or:

$$\text{Erosion} = f \text{ (erosivity, erodibility)}$$

For given soil and vegetation conditions one storm can be compared quantitatively with another and a numerical scale of values of erosivity can be created. Erodibility is similarly quantifiable because the physical and chemical composition of soil controls its resistance to erosion, because measured soil properties can be related statistically to the sediment removed by a particular process, and because the shear strength of soil can be determined for various moisture contents.

7.2a Erosivity

The effectiveness of erosion processes on a slope is closely related to the disposition of water on or in the soil. A number of alternative routes for the water can be visualized as follows:

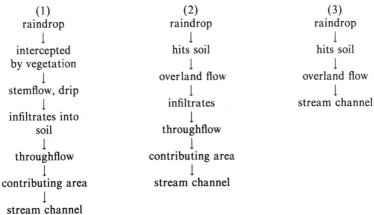

In a fully vegetated catchment it is rare for overland flow to occur except along the edges of stream channels where water, which has infiltrated and then moved vertically and laterally through the soil profile, reaches the saturated zone and emerges at the surface as stormflow (Fig. 7.2). The number of quantitative studies which have verified this model are few, but those of Betson (1964), Whipkey (1965), Ragan (1968) and Dunne (1970) support it for humid temperate forests. The implications of this partial-area model are very great, for it follows that surface wash is likely to be relatively slight and the supply of sediment to a stream must come from channel bank collapse and, more rarely, from minor sources such as subsurface pipes, animal burrows or, where they occur, landslide scars. The routing of most drainage waters through the soil helps to explain why in excess of 50 per cent of the denudation of most forested catchments is by solution (Meade, 1969).

The infiltration capacity of many soils, especially those on steep slopes, is very variable, and a raindrop or thread of water which has moved down a plant stem may reach an impermeable patch of soil and move over the soil or litter layer before reaching a more permeable patch or crack into which it can infiltrate. Reported rainfall intensities reach little more than 225 mm hr^{-1} but infiltration capacities range from 2 to 2500 mm hr^{-1} with common values ranging from 5 to 150 mm hr^{-1} (Hudson, 1971; Morgan, 1969; Selby, 1970). It is, therefore, only under extreme storm events that overland flow is likely to occur under natural conditions. Bare arable soils frequently have relatively low infiltration rates because of compaction of the soil by machinery, so that overland flow is relatively common on croplands when the soils are exposed.

The overland flow model of runoff was developed by Horton (1933, 1938, 1945) who postulated that overland flow would occur as soon as rainfall intensity exceeded the infiltration rate. This model appears to be particularly applicable to areas of bare rock, to arid and semiarid climates, to some arable soils, to frozen soils, and to periods of

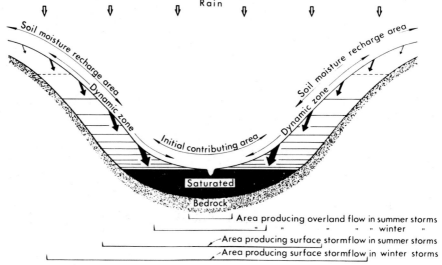

FIG. 7.2 THE DYNAMIC AREA CONCEPT OF RUNOFF PRODUCTION AS DEVELOPED BY THE TENNESSEE VALLEY AUTHORITY (1964)

snowmelt. It is not generally applicable to runoff from vegetated surfaces under humid climates. Where it does occur overland flow has relatively low kinetic energy for the entrainment of sediment compared with the energy of large falling raindrops (Hudson, 1971, p. 62). Most of the erosion occurring on bare soil surfaces results from impacting drops breaking down soil aggregates, splashing them into the air, and causing turbulence in surface runoff.

Erosivity under a vegetation cover is thus low, and from catchments with a full forest or grass cover erosion takes place largely in solution and from the few areas of exposed soil in banks, terracettes and animal burrows. Where flow can become channelled, as for

FIG. 7.3 RILL AND GULLY EROSION ON LONG SLOPES IN THE SOUTH ISLAND, NEW ZEALAND
The attempts to control erosion by planting a few belts of trees have been ineffective because they have not covered the areas contributing water to the channels, and they need to be accompanied by measures to prevent sediment entrainment within the gullies. (Photograph: Ministry of Works and Development, New Zealand)

example when vegetation cover is removed from slopes, then the erosion rate can be vastly increased as rills and gullies cut through the protective root network and entrain the less cohesive subsoil (Fig. 7.3).

7.2b Erodibility

The soil characteristics which control erodibility are particle size distribution, structure, organic matter content, permeability, and root content (Wischmeier, Johnson and Cross, 1971). These variables are of importance because they control the infiltration capacity and the cohesive and frictional strength of the soil. Root content and organic matter content are contributors to apparent soil cohesion. Roots alone can increase effective soil shear strength by two or three times (O'Loughlin, 1974; Waldron, 1977) and this is significant when considering the implications of man-induced changes of vegetation and land use.

7.3   LANDSLIDING

The mechanisms and causes of landsliding have been discussed at considerable length in the literature (e.g. Terzaghi, 1950; Carson and Kirkby, 1972). Discussion here will consequently be confined to the forces resisting and promoting landslides, to the implications for rural and urban land use, to the periodicity of landsliding, and to its effects on valley floors.

7.3a Shallow translational landslides

By far the most common forms of landslides are shallow features which are classified as debris slides, debris avalanches or debris flows depending upon the degree of remoulding of the soil material and the water content of the sliding soil. The stability of a hillslope against landsliding may be expressed as a factor of safety (F) where

$$F = \frac{\text{the sum of resisting forces}}{\text{the sum of driving forces}}$$

Where slope failure is likely to produce a shallow planar slide of great length and considerable width compared with the depth, an infinite slope analysis is applicable (Skempton and De Lory, 1957). Such an analysis ignores possible lateral effects.

The resisting forces are the effective shear strength of the soil and the driving forces are the gravitational forces tending to move the soil down the hillslope, together with the tendency of water to reduce soil strength and to cause a bouyancy effect within the soil. The significance of the gravitational forces will increase as the slope angle increases.

The effective shear strength of the soil is given by the Coulomb equation as:

$$s' = c' + (\sigma - u) \tan \phi'$$

where s' is the effective shear strength at any point in the soil;

   c' is the effective cohesion as reduced by loss of surface tension;

   σ is the normal force imposed by the weight of solids and water above the point in the soil;

u is the pore water pressure derived from the unit weight of water (γ_w) and the hydrostatic head ($\gamma_w\, m\, z$) (see Fig. 7.4)

ϕ' is the angle of friction with respect to effective stresses.

Substituting for σ, the Coulomb equation can be rewritten as:

$$s' = c' + (\gamma\, z\, \cos^2\beta - u)\, \tan\phi'$$

where γ is the unit weight of soil;

z is vertical depth of soil above the failure plane;

β is the angle of inclination of the slope.

As the driving forces are $\gamma\, z\, \cos\beta\sin\beta$ then:

$$\text{Factor of safety} = F = \frac{c' + (\gamma\, z\, \cos^2\beta - u)\, \tan\phi'}{\gamma\, z\, \sin\beta\, \cos\beta}$$

It is convenient to express the vertical height of the water table above the slide plane as a fraction of the soil thickness above the plane, and this is denoted by m. Then if the water table is at the ground surface $m = 1.0$, and if it is just below the slide plane $m = 0$. Pore water pressure on the slide plane assuming seepage parallel to the slope, is given by:

$$u = \gamma_w\, m\, z\, \cos^2\beta$$

thus $$F = \frac{c' + (\gamma - m\, \gamma_w)z\, \cos^2\beta\tan\phi'}{\gamma\, z\, \sin\beta\, \cos\beta}$$

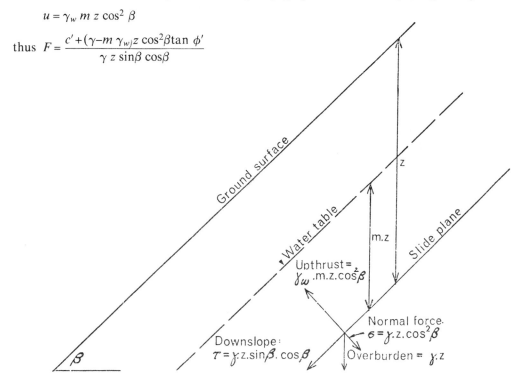

Net normal component = $(\gamma.z - \gamma_w.m.z)\cos^2\beta$

FIG. 7.4 STRESSES WITHIN A SOIL MASS ON A SLOPE

Given that:

$\phi' = 12°$; $c' = 11.9$ kN/m²; $\gamma = 17$ kN/m³; $\beta = 15°$; $z = 6$ m; $m = 0.8$; $\gamma_w = 9.81$ kN/m³
then

$$F = \frac{11.9 + (17 - 0.8 \times 9.81)\, 6 \times 0.92 \times 0.2}{17 \times 6 \times 0.25 \times 0.96} = 0.9$$

Thus the slope is prone to long term failure when the water table is about a metre below the ground surface. If the water table can be lowered by drainage to just below the slide plane (when $m = 0$) then $F = 1.3$ and the slope will be stable against shallow landsliding.

In extremely dry summer periods soils frequently crack and, depending upon the shrinkage limits of the soil, these cracks may extend to depths of a metre or more. In both California and the North Island of New Zealand severe summer rainstorms frequently fall upon dry catchments with deeply cracked soils. As a result water rapidly fills the cracks and seeps downslope under the hydrostatic pressure of water filling the crack. The effective piezometric surface is then well above the soil surface ($m = 1.3$ to 1.6). Such storms frequently give rise to landsliding where rainfall intensities are very high and are in excess of the infiltration capacity of the soil at the base of the tension cracks (Selby, 1967, 1976; Campbell, 1975).

The components of the infinite slope analysis show that only two variables can be readily modified by human influence to increase slope stability, namely water content of the soil and apparent cohesion. Water content can be modified by drainage programmes and apparent cohesion can be modified in so far as it is influenced by the root content of the soil and therefore by the vegetation or land use cover.

7.3b Effects of vegetation upon slope stability

The root networks of all plants provide mechanical reinforcement of the soil and in shear box testing of soil strength this is apparent as an increase in cohesion (Table 7.2). Some roots grow downwards through potential failure zones of shallow landslides into underlying soil or rock, but in shallow soils most trees have shallow root networks which interlock. This interlocking strength is very variable although it usually falls in the range of 1.0 to 12.0 kPa. Trees, but not smaller plants, have three other effects on slope stability (Brown

Table 7.2
STRENGTH ADDED TO SOIL BY PLANT ROOTS

Plant	Soil	Increase in apparent cohesion (kPa)
Conifers (pine, fir)	glacial till	0.9—4.4
Alder	silt loam	2.0—12.0
Birch	silt loam	1.5—9.0
Podocarps	silty gravel	6.0—12.0
Alfalfa (lucerne)	silty clay loam	4.9—9.8
Barley	silty clay loam	1.0—2.5

Note: in any soil the strengthening effect of roots varies greatly with depth and laterally.

Source: Unpublished material; and O'Loughlin, 1974; and Waldron, 1977

and Sheu, 1975; Gray, 1970). First, wind throwing and root wedging occurs as trees are overthrown by strong winds or under heavy snow falls. Secondly, trees increase the surcharge on a soil mantle and have the effect of increasing the normal load upon the failure plane and thus of increasing slope stability – but this is only a significant effect when large, closely spaced trees occur on thin soils and provide an increase in normal stress of up to 5 kPa, while increasing shearing stress by only half that (Bishop and Stevens, 1964). Thirdly, trees may also increase evapotranspiration from the soil and so lower water tables.

The importance of trees for slope stability is demonstrated in many areas where deforestation has been followed, after an interval of a few years, by periods of severe shallow landsliding. Examples of this sequence are known for New Zealand, Alaska, British Columbia, the Himalayan foothills, and Japan. The interval which occurs between clearance and landsliding may be due to the lack of a storm with sufficient intensity and duration to raise porewater pressures to critical levels, but it is more commonly attributable to the slow decay of the tree roots which are left in the soil. The thinnest roots decay first so that over the first year or so after clearing, root tensile strength per unit volume of soil decreases rapidly and then over the following two to ten years the larger roots decay until they contribute nothing to soil apparent strength. The actual rate

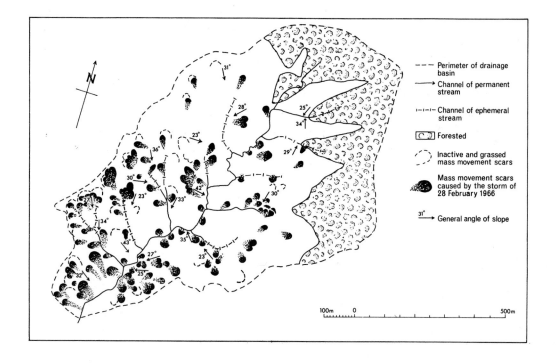

FIG. 7.5 THE DISTRIBUTION OF SHALLOW TRANSLATIONAL LANDSLIDES IN ONE SMALL CATCHMENT IN THE HAPUAKOHE RANGE, NEW ZEALAND
The slides developed during a storm on 28 February 1968, when nearly 250 mm of rain fell in twenty-four hours. Landsliding was confined to areas with pasture grass cover and the forested area was not affected.

of decay depends upon the tree species, the soil climate, and the rooting pattern of the trees.

The effect of trees upon slope stability is particularly noticeable where adjacent tree-covered and grass-covered slopes are subjected to the same storm. The incidence of landslides is nearly always far greater on the grass-covered slopes (Selby, 1967; Pain, 1971; Swanston, 1970) as illustrated in the basin shown in Figure 7.5.

7.3c Effects of landsliding on valley floors

The debris from very large individual landslides or that from many small slides occurring during a single storm over a catchment can have effects on valley floors which may be evident for hundreds of years (Figs. 7.6, 7.7). Large slumps and flows in rocks, such as mudstones, may infill valleys to depths of many metres and, once the vegetation cover has been disrupted by a slide, continued erosion by gullies may provide so much sediment that streams are incapable of removing it. This has happened in parts of Japan (Machida, 1966) and in northern New Zealand where a series of severe rainstorms in the 1930s triggered numerous landslides. In the Waipaoa River catchment some valley floors have been infilled to depths of over 30 m since then and aggradation is still continuing in spite of attempts to afforest the slopes.

The northern Ruahine Range (Fig. 7.7), where the underlying bedrock is severely

FIG. 7.6 A MASSIVE LANDSLIDE IN MUDSTONES
This has produced a tongue of remoulded debris which moved down the valley for nearly a kilometre. It has blocked the tributary valleys and infilled the valley floor. Southern Hawke's Bay, New Zealand.
(Photograph: J. Pettinga)

FIG. 7.7 A VALLEY FLOOR INFILLED WITH ROCK DEBRIS DERIVED FROM NUMEROUS SHALLOW LANDSLIDES IN THE
NORTHERN RUAHINE RANGE, NEW ZEALAND

shattered sandstones and argillites, has been subjected to a series of severe storm events
over the last 400 years which have left behind a sequence of terraces (Grant, 1965). The
approximate age of each terrace has been determined from the oldest trees on it and from
the known age of volcanic ash deposits which have fallen onto each terrace surface; the
basic assumption being made that the terrace cannot be younger than the trees or ashes
on it. Terraces are dated as being pre-130, 1450, 1650 and 1840 A.D. The more recent
terraces suggest that very large catastrophic events may occur every 200 years or so, but
it is certain that more frequent severe storms also cause much erosion. The debris of lesser
events may, however, be largely removed from catchments either in a series of pulses as
large floods carry waves of debris down the channels, or by minor floods which occur
several times a year.

It can be seen from Figure 7.7 that near the headwaters of the streams large influxes of debris bury the channel forms and low terraces. The debris in this channel was all deposited in one major event and the minor terraces alongside the channel were formed in the two years afterwards. The datable evidence for severe storm periods is consequently preserved not on the slopes or in the channels of the headwaters but in the middle reaches of the rivers where terraces are not necessarily buried by the subsequent wave of debris so that terrace remnants are preserved, at least discontinuously, along reaches of migrating channels.

Catastrophic slope erosion and infilling of valley floors is not confined to Japan and New Zealand. It has been described from the Caucasus and from California (Starkel, 1976; Lamarche, 1968). In the Caucasus the term *sjel* is applied to rapid mudflows and debris-laden streams with very high densities. Mudflows with densities of $1 \cdot 7$ to $2 \cdot 3$ tonnes m^{-3}, are capable of transporting very large boulders because the boulders have densities of $2 \cdot 5$ to $2 \cdot 6$ t m^{-3} and virtually float in the mud. Several sjels may occur in one valley in a single year.

The effects of landslides on valley floors can be of even greater economic importance than the landslides themselves. In steep hill country, roads, railways, and power and telephone cables frequently follow the valley floors and may be buried by debris. Perhaps of wider significance is the general raising of stream beds so that channels become unstable, flood protection banks are overtopped, bridges are buried or their clearance above flood levels is reduced, and the agricultural production from floodplains is lost or impaired.

Very large landslides of the rock slide and rock avalanche type which are reported from high mountain ranges such as the Canadian Rockies (Cruden, 1976), the Peruvian Andes (Browning, 1973) and the Himalayas (Kingdom-Ward, 1955) have an even more devastating effect for they are capable of burying towns and blocking valleys. The lakes impounded behind the landslide debris may eventually burst through the temporary dam to release a devastating flood wave upon the valley downstream.

7.4   PERIODICITY OF EXTREME EVENTS ON SLOPES

Severe storms with heavy rainfall are the most effective agents of change on hillslopes. The greater the magnitude, that is the amount and intensity, of the rainfall, the lower is its frequency. Very severe storms, therefore, occur separated by long intervals of time but the frequency with which land-forming processes are effective depends partly upon the type of process and partly upon the climate of a region. In any area it must also be affected by the permeability of the regolith and the closeness of the hillslopes to limiting angles of stability.

It has been contended that in rivers most of the work of transportation is carried out by floods which are of such magnitude that they recur with a frequency of at least once in five years (Wolman and Miller, 1960). On hillslopes, processes such as solution occur during every prolonged rainfall but the processes which remove large quantities of debris, such as gullying and landsliding, occur at infrequent intervals. The frequency of these catastrophic events has a major effect upon the rate at which the surface of the earth changes.

Hillslope processes which are active frequently do not greatly disturb the approximate balance which usually exists between the rate of weathering and the removal of regolith. Except in deserts, therefore, there is an approximate equilibrium between weathering, soil development, the vegetation cover, and the rate of erosion on hillslopes. Extreme events destroy this equilibrium, breaking the vegetation cover, stripping away regolith along the track of gullies and landslides, and producing large influxes of sediment into valley floors. As it is uncommon for several extreme events to occur in a short interval, the scars and the deposits may then be slowly modified by lesser intensity processes such as creep, solution, and wash, while weathering, soil formation and vegetation gradually re-establish a surface which is in approximate equilibrium with the energy of the processes usually acting on the slope. After a period the equilibrium may be broken by another extreme storm. This type of episodic development of the land surface can be visualized as occurring in a step-wise manner (Fig. 7.8) in which storm events are followed by gradual periods of adjustment to the normally active processes and then, when an approximate equilibrium is re-established, by a period of relative stability.

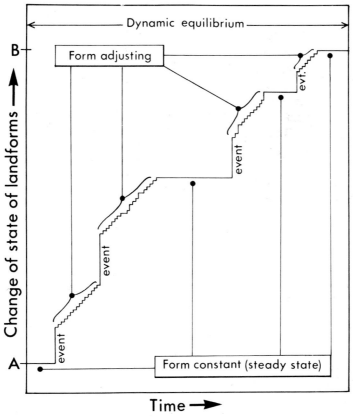

FIG. 7.8 IMPLICATIONS OF DYNAMIC EQUILIBRIUM
Within the term dynamic equilibrium there are subsumed the three states of: (i) a landforming event; (ii) the adjustment of form which results after that event; and (iii) a period of steady state in which there is no adjustment of form. (Source: Selby, 1974a)

The importance of extreme events in different regions of the world is very variable. Starkel (1976) reviewed studies of extreme events and concluded that four classes of region may be distinguished.

1 Regions with a frequency of extreme events of 5–10 per century and with events of such magnitude that in each the denudation greatly exceeds the denudation by all low-intensity processes during 100 years. Such areas are most common in tropical monsoon and mediterranean climates, and in steep uplands and farming lands of the temperate zone where human action has removed or changed the vegetation cover.

2 Regions in which extreme events are rare and in which such events do not exceed the total denudation of 100 years of low intensity processes. In these areas heavy rainfalls or snowmelts occur each year. Such regions are common in the semiarid zone.

3 Regions with very rare extreme events. Because of normally very low precipitation a storm may achieve much more denudation than usually occurs in 100 years. Arid zones and some parts of the boreal zone are in this category.

4 Stable regions showing little variation from normal denudation rates. Many Arctic, Antarctic, continental boreal and lowland zones of the temperate regions are in this group.

In any part of the world, mountains are far more subject to extreme events than lowlands, but even in them the effects of extreme denudation are frequently limited to quite small localities. Thus catastrophic events may occur in a mountain range nearly every year but each time hitting a different area so that overall the frequency is much less than one event a year. Similarly the sediment produced by a storm event may range from a few tonnes km^{-2} to hundreds of thousands of t km^{-2}, and the average downwearing of the affected areas from 0·01 to 200 mm. This range occurs because the area of the ground suffering extreme erosion during a storm may vary from less than 1 to more than 50 per cent.

Data on the frequency of extreme events are not available for many parts of the world, but some comparisons are presented in Figure 7.9. In the hills and lowlands of Western Europe periods of landsliding appear to be related to intervals of wet climate since the last

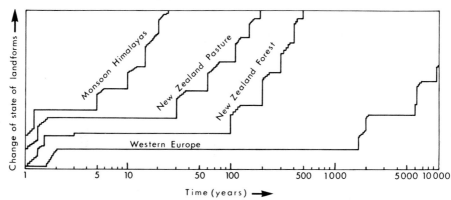

FIG. 7.9 CURVES SHOWING THE MAGNITUDE AND FREQUENCY OF LANDFORM CHANGE BY LANDSLIDING FOR SELECTED ENVIRONMENTS
(Source: Selby, 1974a)

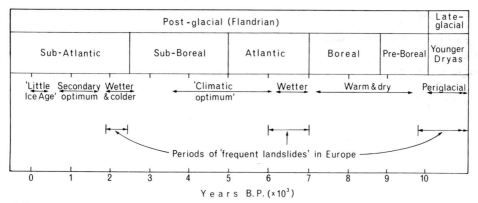

FIG. 7.10 PERIODS OF LANDSLIDING IN LOWLAND WESTERN EUROPE DURING THE LAST 11 000 YEARS (Source: Starkel, 1966)

glacial (Fig. 7.10) (Starkel, 1966). These periods last for a few hundred years and occur at intervals of 2000 years. At the other end of the scale the frontal ranges of the Himalayas north of the Ganges delta receive prolonged and intense rainfalls nearly every year and extreme landforming events probably have a frequency of about once in five years (Starkel, 1972).

Less frequent are the storms causing landslides in the North Island of New Zealand. Under original forest the hills are affected by shallow landsliding about once in every 100 years, but in the last century many uplands have been cleared of forest, and pastures now cover the hills. Pastures are less protective than the forests, and storms of lower intensity, which recur about once in thirty years, now cause landslides (Selby, 1967, 1976; Pain, 1969).

7.5 SLOPES IN URBAN AREAS

In densely populated areas the spread of settlement on to steep slopes may be necessary, and it is often desired by people seeking extensive views or freedom from the air pollution of enclosed valleys. On a slope which has a factor of safety close to unity, almost any increase in the driving forces or decrease in resisting forces can be hazardous. When roads and buildings are placed on slopes, or sewers, water pipes, and power cables are buried in them there is nearly always an accompanying change in the stresses acting on the slope materials. With good planning the probability of instability developing, as a result of stress changes, can be diminished; but with bad planning, or no planning, the probability can be increased.

Slope failure may be encouraged by many results of construction works but two features are particularly important: first, the removal of the toes of slopes reduces lateral support for the soil above; and secondly, the arrival of extra water is a common triggering event for landslides. The removal of the toes of slopes is frequently necessary if roads are to be built across the slope. Where soils are thin and the road can be cut into firm rock there may be no hazard, but deep soils with downslope water seepage may be severely affected.

An infinite slope analysis may indicate whether the slope is close to limiting equilibrium but before detailed planning commences many potential hazards can be recognized by careful study of aerial photographs and soil conditions. The best indications of future instability are their existence in the past, for many landslides occur within the site of old slides simply because landslide debris usually has a significantly lower shear strength than undisturbed soil, and the drainage of debris from old slides is usually impeded. Signs of instability include cracked or hummocky surfaces, crescent-shaped depressions, crooked fences, trees or lamp-posts leaning uphill or downhill, uneven road surfaces, swamps or wet ground in elevated positions, rushes or other wet-site vegetation growing on slopes, and water seeping from the ground.

The arrival of excess water at a site may result from badly designed storm-water drains or it may be seepage from leaking swimming pools, the irrigation of lawns, runoff from roofs and sealed areas, or the diversion of natural drainage.

All surface investigations should include measurement of slope angles of adjacent stable and unstable areas with similar geology, groundwater conditions and vegetation cover to that of the area which is to be developed. This information provides a useful guide to the likely stability of slopes before and after construction. Such preliminary investigations then have to be followed by detailed investigations of rock, soil, and groundwater conditions.

From suitable information it is possible to determine which slopes can be safely traversed by roads or be cut into for house sites, and which areas need special reinforcement of cut slopes and diversions of drainage.

Slope stability investigations and protection works are often expensive but they are almost invariably cheaper than the remedial works which are necessary if a slope does fail in an urban area. Precautionary measures taken before development also have the advantage that they can be charged to the intending occupier, but remedial works are often charged to the community, either through local taxes or increased insurance premiums.

7.6   DENUDATION AND LAND-USE PLANNING

The four main variables which influence the rates of downwearing of the land surface – climate, rock type, vegetation and slope angle – themselves contain such wide variations that it is probably premature to suggest that we have anything better than an order of magnitude estimate and understanding of the rate of denudation. Knowledge of actual rates is, however, essential for informed planning of land use. In most parts of the world it is probably true that lowland and gentle hillslope surfaces are worn down at rates of the order of 50 mm/1000 years, and mountain slopes are worn down at rates of 500 mm/1000 years (see Selby, 1974b; Young, 1974 for reviews of the evidence). Agricultural, deforestation, and construction activities by man have probably increased world denudation rates two or three times above the long-term geological rate and locally the accelerated rates may be 1000 times as great.

An appreciation of the effect of vegetation cover on erosion rates has been obtained from the work of the United States Soil Conservation Service (see Chapter 12) but it is still unclear how the partial-area concept of contributing areas for stormflow, and also understanding of the periodicity of landsliding affect currently available estimates of denudation. Many measurements of processes have been made on slope units which are not part of the dynamic zone of water and sediment supply to streams and hence these measurements may not be representative of slope processes.

The partial-area model has provided a valuable indication of where soil conservation works need to be located for maximum effect. Conservation works such as tree planting and the formation of thick grass strips are clearly best positioned in the stormflow contributing areas around a channel. In areas where overland flow can occur the primary aim must be to reduce the depth and length of flow over the slope by employing strips of closely spaced vegetation or shallow grassed channels along the contour (see Chapter 12).

Landslide hazards are far more difficult to prevent. Extremely vulnerable areas can be planted, or even contained by retaining walls if this is warranted, but appreciation of the probable frequency of landsliding allows a comparison to be made between an economic assessment of the cost of repair works and the cost of protection works. Whether or not expensive protection works are erected depends, of course, not only on cost comparisons but on the social and political acceptance of dangers to life and property. Whether or not the knowledge of slope processes now available is used for better planning in rural and urban areas depends partly upon supplying expertise to decision makers but, at least as importantly, it depends upon the political will to use that information.

REFERENCES

BETSON, R. P., 1964, 'What is watershed runoff?' *Jnl. Geophys. Res.*, 69, pp. 1541–51.

BISHOP, D. M., and STEVENS, M. E., 1964, 'Landslides in logged areas in SE Alaska.' *US Dept. of Agric., Forest Service Res. Pap.*, NOR-1, pp. 1–18.

BROWN, C. B., and SHEU, M. S., 1975, 'Effects of deforestation on slopes.' *Jnl. of the Geotechnical Engineering Division, Amer. Soc. of Civil Engineers*, GT2, pp. 147–65.

BROWNING, J. M., 1973, 'Catastrophic rock slide, Mount Huascaran, north-central Peru, May 31, 1970.' *Amer. Assoc. of Petroleum Geologists Bull.*, 57, pp. 1335–41.

CAMPBELL, R. H., 1975, 'Soil slips, debris flows, and rainstorms in the Santa Monica Mountains and vicinity, southern California.' *US Geol. Surv. Prof. Paper*, 851, pp. 1–51.

CARSON, M. A., and KIRKBY, M. J., 1972, *Hillslope Form and Process* (CUP, Cambridge).

CRUDEN, D. M., 1976, 'Major rock slides in the Rockies.' *Can. Geotech. Jnl.*, 13, pp. 8–20.

DEARMAN, W. R., 1974, 'Weathering classification in the characterisation of rock for engineering purposes in British practice.' *Bull. Int. Assoc. of Engineering Geol.*, 9, pp. 33–42.

DUNNE, T., 1970, 'Runoff production in a humid area.' *US Dept. of Agriculture*, ARS 41–160, pp. 1–108.

GRANT, P. J., 1965, 'Major regime changes of the Tukituki River, Hawke's Bay, since about 1650 A.D.' *Jnl. Hyd. (N.Z.)*, 4, pp. 17–30.

GRAY, D. H., 1970, 'Effects of forest clear cutting on the stability of natural slopes.' *Bull. Assoc. of Engineering Geologists*, 7, pp. 45–66.

HORTON, R. E., 1933, 'The role of infiltration in the hydrologic cycle.' *Amer. Geophys. Union Transactions*, 14, pp. 446–60.

1938, 'The interpretation and application of runoff plot experiments with reference to soil erosion problems.' *Soil Science Society of America Proceedings*, 3, pp. 340–9.

1945, 'Erosional development of streams and their drainage basins.' *Geol. Soc. of America Bull.*, 56, pp. 275–370.

Hudson, N., 1971, *Soil Conservation* (Batsford, London).

Kingdom-Ward, F., 1955, 'Aftermath of the great Assam earthquake of 1950.' *Geog. Jnl.*, 121, pp. 290–303.

Lamarche, V. G., 1968, 'Rates of slope degradation as determined from botanical evidence, White Mountains, California.' *US Geol. Surv., Prof. Paper*, 352I, pp. 341–77.

Machida, H., 1966, 'Rapid erosional development of mountain slopes and valleys caused by large landslides in Japan.' *Geographical Reports of Tokyo Metropolitan University*, 1, pp. 55–78.

Meade, R. H., 1969, 'Errors in using modern stream-load data to estimate natural rates of denudation.' *Geol. Soc. of America Bull.*, 80, pp. 1265–74.

Morgan, M. A., 1969, 'Overland flow and man', *Water, Earth and Man*, ed. R. J. Chorley (Methuen, London), pp. 239–55.

O'Loughlin, C., 1974, 'The effects of timber removal on the stability of forest soils.' *Jnl. Hyd. (N.Z.)*, pp. 121–34.

Pain, C. F. 1969, 'The effect of some environmental factors on rapid mass movement in the Hunua Ranges, New Zealand.' *Earth Science Journal*, 3, pp. 101–7.

1971, 'Rapid mass movement under forest and grass in the Hunua Ranges, New Zealand.' *Austral. Geog. Studies*, 9, pp. 77–84.

Ragan, R. M., 1968, 'An experimental investigation of partial area contributions.' *Int. Assoc. for Sci. Hyd., Berne Symposium*, 76, pp. 241–51.

Selby, M. J., 1967, 'Geomorphology of the greywacke ranges bordering the Lower Waikato Basin.' *Earth Science Journal*, 1, 37–58.

1970, 'Design of a hand-portable rainfall-simulating infiltrometer with trail results from the Otutira Catchment.' *Jnl. Hyd. (NZ)*, 9, pp. 117–132.

1974a, 'Dominant geomorphic events in landform evolution.' *Bull. Int. Assoc. of Engineering Geol.*, 9, pp. 85–9.

1974b, 'Rates of denudation.' *NZ Jnl. Geog.*, 56, pp. 1–13.

1976, 'Slope erosion due to extreme rainfall: a case study from New Zealand.' *Geografiska Annaler*, 58A, pp. 131–8.

Skempton, A. W., and De Lory, F. A., 1957, 'Stability of natural slopes in London Clay.' *Proc. Fourth Int. Conf. on Soil Mechanics and Foundation Engineering, London*, 2, pp. 378–81.

Starkel, L., 1966, 'Post-glacial climate and the moulding of European relief.' *Proc. Int. Symp. on World Climate 8000 to 0 B.C.*, Roy. Met. Soc., London, pp. 15–32.

1972, 'The role of catastrophic rainfall in the shaping of the relief of the Lower Himalaya (Darjeeling Hills).' *Geographica Polonica*, 21, pp. 103–47.

1976, 'The role of extreme (catastrophic) meteorological events in contemporary evolution of slopes', *Geomorphology and Climate*, ed. E. Derbyshire (Wiley, London), pp. 203–46.

Swanston, D. N., 1970, 'Mechanics of debris avalanching in shallow till soils of south-east Alaska.' *USDA Forest Service Res. Pap.*, PNW–103, pp. 1–17.

Tennessee Valley Authority, 1964, 'Bradshaw Creek-Elk River, a pilot study in area-stream factor correlation.' *Research Paper*, 4, (TVA) pp. 1–64.

Terzaghi, K., 1950, 'Mechanism of landslides.' *Bull. Geol. Soc. of America, Berkey Volume*, pp. 83–122.

Waldron, L. J., 1977, 'The shear resistance of root-permeated homogeneous and stratified soil.' *Soil Science Society of America Jnl.*, 41, pp. 843–9.

Whipkey, R. Z., 1965, 'Subsurface stormflow from forested slopes.' *Int. Assoc. for Sci. Hyd., Bull.*, 10(2), pp. 74–85.

Wischmeier, W. H., Johnson, C. B., and Cross, B. V., 1971, 'A soil erodibility nomograph for farmland and construction sites.' *Jnl. Soil and Water Conservation*, 26, pp. 189–93.

Wolman, M. G., and Miller, J. P., 1960, 'Magnitude and frequency of forces in geomorphic processes.' *Jnl. Geol.*, 68, pp. 54–74.

Young, A., 1974, 'The rate of slope retreat.' *Inst. Brit. Geog. Special Publ.*, pp. 65–78.

8

River Channels

K. J. GREGORY
University of Southampton

It is with rivers as it is with people: the greatest are not always the most agreeable nor the best to live with.

Henry van Dyne (1852–1933)

8.1 INTRODUCTION

The impact of man on river channels can be found in the extent to which the river has been *agreeable*; the way in which rivers have been *lived with*; the changes which man's activities have occasioned in river channels; and the ways in which man has deliberately endeavoured to engineer river channels to ensure that the river is more agreeable and easy to live with.

Man's impact on river channels is not confined to the last few decades. This is underlined in China where rivers are of great significance and where the Yangtze is regarded as 'China's main street' because of its potential for navigation and the Hwang Ho, or Yellow River, is thought of as 'China's sorrow' because of its history of flooding, its shifting channels, and its high silt load. The dimensions of the problem are indicated when it is recalled that a quarter of the Hwang Ho's drainage basin is able to provide loess to contribute to the high silt load which represents up to 46 per cent, by weight, of the flow in the river channel; that in a century the Hwang Ho carries sediment equal to a layer 3 cm thick from the entire basin; that the river has flooded and broken through levees many times during the last 4000 years; and that during the last 4262 years there have been twenty changes of the river's course. Some of these changes have been very substantial and the major shifts are illustrated in Figure 8.1. Because of the high population of 110 million within the drainage basin (the quarter of a million square kilometres that were susceptible to flooding), the impact of major events like 1855 when 250 000 people died, and in addition the characteristics of the river itself, the river channel

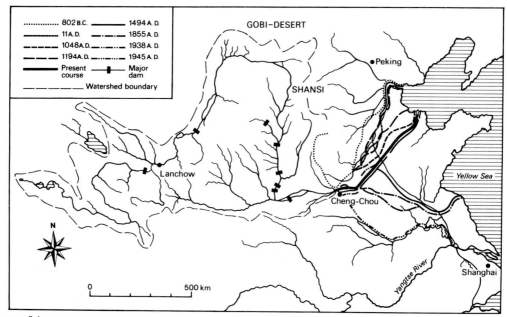

FIG. 8.1 MAJOR CHANGES OF THE COURSE OF THE YELLOW RIVER (HWANG-HO) CHINA
The major changes of course illustrate the major shifts that can occur and the diversion to the south-east was deliberately effected during war time conditions. The location of major dams is shown. (Source: Tsung-lien Chou, 1976)

has been the subject of control particularly since Liberation. Since then irrigation work and flood prevention have been further developed so that 1800 km of restraining dyke have been built, flow velocities have been reduced by the construction of training walls, piers, and baffles, conservation methods have been extended in the loess areas, and many large dams have been constructed. Six large dams exist on the main river (Fig. 8.1) and the Sanmen gorge dam built between 1956 and 1960 could control 92 per cent of the water, but siltation in the first four years of operation had reduced the reservoir capacity by 50 per cent.

Although developed more recently, control of the Mississippi has also been very extensive. The first levees were built in 1699 by individual land owners, by 1844 levees were almost continuous from 30 km below New Orleans to the mouth of the Arkansas river on the west bank and to Baton Rouge on the east bank of the river, and in 1879 Congress founded the Mississippi River Commission. Regulation over the succeeding century led to the gradual control of floods – up to 1931 there was a flood every 2·8 years, flooding occurred every 2·2 years between 1932 and 1943, but only once every 5·5 years between 1944 and 1955, and there was only one overbank condition between 1956 and 1969 (Stevens, Simons and Schumm, 1975).

These two examples illustrate the need to control rivers as man's activities have increasingly extended within the rivers' domain, and so we need to establish the ways in which man has modified processes in river channels (8·2), to understand the effects that these modified processes have had (8·3), and to know of the alternative strategies for further developments (8·4).

8.2 DOWN THE RIVER

R. L. Sherlock, who wrote *Man as a Geological Agent* in 1922, noted that when Queen Victoria came to the throne, the River Tyne in north-east England was a tortuous, shallow river, full of sand banks and eddies, and it was fordable whereas in 1922 the river was severely affected by industry and was regulated. Fifty-six years later in a paper on world dams, Beaumont (1978) noted that 20 per cent of the total runoff in Africa and North America is regulated by reservoirs, 15 per cent in Europe, and 14 per cent in Asia. There was a surge of dam building between 1900 and 1940 and a marked rise occurred after 1950. Between 1922 and 1978, the dates of these two works, many changes of river channel processes occurred and this was accompanied by a growing awareness of the effects which these changes of process (Chapter 5) could have upon river channels.

8.2a Direct modifications of processes

Processes in river channels have been modified directly and indirectly as a result of human influence. Direct modifications involve those cases where the river channel is modified in some way and Mrowka (1974) cited dam and reservoir construction, channelization, bank manipulation and levee construction, and irrigation diversions as examples of direct channel manipulation. Such changes can be resolved into two major categories. The first category consists of those changes which arise consequent upon a change of the river channel at a specific location. This includes the construction of a dam to impound a reservoir for water supply, for flow regulation, or for the generation of hydroelectric power, the abstraction of river flow for water supply or for diversion to another river system, the return of water at a specific outfall point from industrial use or from water supply systems, and also the building of structures which intersect the river channel at a specific point, such as the bridges needed by road and rail transport networks. In all these cases the water and sediment in the river may be modified. Thus, the construction of a dam provides a storage reservoir upstream and this will modify the pattern of downstream discharges (Rutter and Engstrom, 1964). If the reservoir is not full then inflow may be contained within the reservoir so that peak discharges downstream may be reduced by 98 per cent, but even if the reservoir is near capacity level the routing of a flood peak through the reservoir can lead to a peak discharge downstream reduced by more than 50 per cent (Moore, 1969). Because storage of water in the reservoir may increase the residence time of water in the river system, then losses by seepage and evaporation can be increased giving a reduced volume of river flow. In addition to the major dams of the world there are many small ones used for local water supplies and conservation measures, so that it is evident that flow regulation is of widespread occurrence. In general the magnitude and frequency of peak flows is reduced downstream from a dam, and below Hoover dam on the Colorado the magnitude of the mean annual flood has been reduced by 60 per cent (Dolan, Howard and Gallenson, 1974). Also on the Colorado, it has been shown that some 9·87 million m^3 of bottom sediment were scoured from the channel downstream of Glen Canyon dam (constructed between 1956 and 1963) although it subsequently became quite stable between 1965 and 1975 (Pemberton, 1976). Because reservoirs provide a sediment trap in which up to 95 per cent of the bedload and suspended sediment carried by the river can be retained, then this encourages scour downstream of the dam. At the upstream end of the reservoir, sediment accumulation may be encouraged as the velocity of the water is reduced on reaching the impounded water body.

Whereas other point modifications may have less dramatic effects on water and sediment discharge, the consequences will depend on the way in which abstraction of water and sediment relates to peak flow rates. If a constant amount of water is abstracted from river flow this will not significantly affect high discharges so that any influence on river regime may be apparent for only a short distance downstream. Similar local effects are likely where bridges are built because water velocity may be increased and sediment supply inhibited at the bridge site.

A second category of directly induced change of river regime arises where a reach of river channel has been modified. This includes those cases where the pattern has been modified by a series of cutoffs, where the channel has been modified by dredging to maintain navigation or to supply gravel, or where the channel has been regulated to reduce flooding, to drain wet areas, to control bank erosion or to improve river alignment (section 8.3). The effect of creating a larger channel, or one which is reduced in roughness, is to increase water velocity so that the frequency of peak discharges may be increased downstream of the modified reach. Thus the Missouri river from Sioux City, Iowa to its confluence with the Mississippi has been reduced in length from 269 km in 1890 to 217 km in 1960 as a result of flood control and navigation improvement (Schumm, 1971; Richardson and Christian, 1976).

Modifications at specific points along a river, and along river reaches, can therefore induce adjustments of the frequency of peak discharges, of the total runoff amount, and of the level and persistence of low flows. In addition, the amount of sediment available for

Table 8.1
EXAMPLES OF MAN-INDUCED CHANGES RELEVANT TO RIVER CHANNEL PROCESSES

Area	Runoff regulated by reservoirs as percentage of total runoff
Europe	15·1
Asia	14·0
Africa	21·0
North America	20·6
South America	4·1
Australasia	6·1

Source: Beaumont (1978)

Of world cultivated land, 15 per cent is irrigated and can lead to changes in river channel processes
In USA to 1968, 15479 km of riverbank had been treated by Corps of Engineers projects involving mattresses, jetties, training wells, revetment, dikes etc

Number of projects	Summary of downstream flood control works in US to 1966 Type	Extent
263	reservoir	total storage 124310 acre feet
426	levee and floodwall	14115 km
193	channel improvement	8578 km

Source: Todd (1970)

transport may be reduced, as for example where a river channel is regulated to inhibit bank erosion, or increased where gravel extraction makes more bed material available or where mine spoil is provided to the channel. In studies in central Wales it has been shown (Lewin, Davies, and Wolfenden, 1977) how the wastes from nineteenth century mining were indiscriminately fed into local streams such as the Ystwyth until preventative legislation was enacted. Further examples of man-induced changes of channel processes are included in Table 8.1.

8.2b Indirect modifications of channel processes

Indirect changes include all those instances where a change of the character of the catchment area is responsible for an alteration of the magnitude and frequency of water and sediment discharge. A variety of land-use effects are now known to have occasioned changes in discharge in sediment hydrographs and in bedload transport (Chapter 5) and these changes either involve an increase or a decrease in the activity of river channel processes. Whereas deforestation may induce greater activity because of increased sediment availability and higher runoff rates, the converse may obtain where runoff is decreased and sediment transport reduced following afforestation or conservation measures. Land-use changes can be responsible for adjustments of river channel processes which in turn affect channel landforms (Gregory, 1977a). Thus the removal of forest in a basin on South Island, New Zealand was shown by O'Loughlin (1970) to have induced more frequent peak discharges and similar increases were found in the Beskid Sadecki Mountains of Poland following land-use change as potato cultivation was increased (Klimek and Trafas, 1972).

Where the temporal sequence of water flow and sediment transported by rivers is altered, then changes in the processes operating within the river channel should occur as a direct consequence. Such changes can be anticipated from equations such as that proposed in 1955 by E. W. Lane as:

$$Qs D \approx Qw S$$
  Qs = bed material load
  D = particle diameter
  Qw = water discharge
  S = stream slope

He deduced (Lane, 1955) the possible implications of six types of change from this equality. Thus, if water was diverted into the system then Qw would increase so that the equality could be maintained only if either Qs or D increased or if S decreased. A similar approach has been available for many years from regime theory. This is based upon the notion that a channel carrying a definite long-term pattern of water and sediment discharge, and subjected to a definite set of constraints that can adjust to the action of the flow, will acquire a definite regime (Blench, 1972). River morphometrics is then defined as the routine of measuring quantities relevant to river regime and morphometric changes may be interpreted from the present regime (Blench, 1972). The principal changes of channel process that may arise from a change in regime will be expressed either in increased scour by erosion of the banks or bed, or in increased aggradation which is reflected in channel sedimentation. The consequences of man-induced changes of river channel processes should be substantial and Beckinsale (1972) has argued that man has been responsible for many of the changes in sediment load of rivers since the Quaternary. However he reminds us that the effects of such changes are never simple; for example deposition on flood plains in western Europe since Roman times must be viewed in relation to a small rise in sea level. Furthermore, initially greater sediment transport following deforestation could have been succeeded by scour and channel incision as sediment sources were exhausted. An additional complication is afforded by the relative significance of climatic fluctuations and the influence of man and these two factors have been compared in the Mediterranean basin by Vita-Finzi (1969). In south-east Wales Crampton (1969) argued that the sediment yield of rivers was increased following the deforestation of Iron Age times but that this was succeeded by a phase when the resulting fan deposits were dissected because sediment yields decreased again and scour of channels occurred.

In addition to the sequence of temporal changes of processes it is also important to remember the significance of threshold levels and of negative feedback mechanisms. Thus a change in process will induce a result which will be expressed in scour or in aggradation only if the change exceeds a critical threshold level. If a channel is in regime then the sediment load may have to be decreased below a critical level before scour will take place and the threshold level will depend upon the shear strength of the bed and bank materials. The time required to achieve a new steady state after disruption, the relaxation time, has been modelled by a rate law using half-life values analogous to those used for radioactive materials and chemical mixtures (Graf, 1977). Because the change to a new steady state may depend upon the occurrence of a large event or flood then a process change may not immediately prompt a response. A further complication to the expected sequence of change may be the incidence of negative feedback. If the sediment transported by a river is decreased substantially then it would be expected that the river bed may be scoured in accordance with the notion embodied in Lane's equation (P.127). However a layer of coarse material may develop on the channel bed (effectively an increase of D) and so resist scouring. This is illustrated by the Colorado river channel downstream of Glen Canyon Dam, because degradation was controlled by the development of an armouring of cobble-sized materials on the bed (Pemberton, 1976); and below the Aswan dam, Hammad (1972) deduced that an armoured condition may develop before an appreciable change of bed slope could occur on the Nile.

8.3   RIVER CHANNEL REACTION

Changes in river channel processes should induce adjustments of river channel and flood plain morphology. A number of studies have explored the potential adjustments which can occur and Strahler (1956) indicated how gullying in the headwaters of a basin could have aggradation downstream as its corollary (Fig. 8.2). The general model has been exemplified specifically by a study of the sedimentation which averaged 1m in the Alcovy river swamps in Georgia (Trimble, 1970), and of the alluvial fills which accumulated in the Orange and Vaal basins of South Africa following over-grazing and burning of the degraded grassveld after 1880 (Butzer, 1971). An equally important study by Wolman (1967a) identified the possible changes in channels after reservoir construction and urbanization and a general model (Wolman, 1967b) was developed for the sequence of changes that would be expected with land use changes in the north-east USA since 1700 (Fig. 8.2).

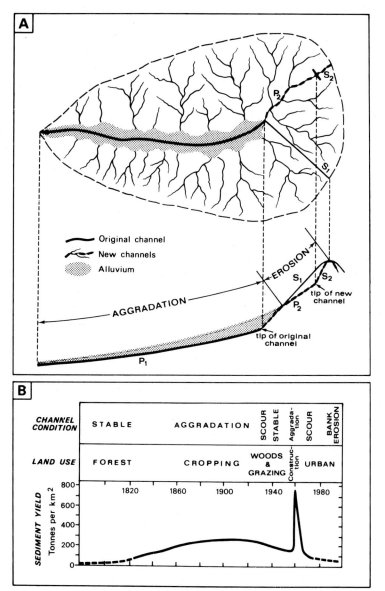

FIG. 8.2 MODELS OF RIVER CHANNEL ADJUSTMENT

Model (A) demonstrates the way in which a change of land use can induce channel extension by gullying and downstream aggradation marked by floodplain growth. (Source: A. N. Strahler, 1956). Model (B) depicts the way in which changes of sediment yield since 1700 in the north east USA (see also Fig 5.7C) are associated with man-induced alterations of land use and induce alterations in river channel stability. (Source: Wolman, 1967b)

Changes of river channels consequent upon changes in river channel processes have been termed river metamorphosis by Schumm (1969, 1971, 1977) and he has provided many illustrations of such metamorphoses. Similarly Richardson and Simons (1976) argued that changes in flow, in either magnitude or time distribution, should be expressed in changes in channel geometry, channel planform and sediment transport. They argued further that river characteristics, including the height of the water surface, depth, and channel geometry, could all respond to changes in sediment transport. There is obviously a range of parameters that can respond to change and Hey (1974) contended that rivers possess five degrees of freedom and later extended this (Hey, 1978) to conclude that it is possible to define bankfull hydraulic geometry by the simultaneous solution of nine process equations (Table 8.2). This approach demonstrates that changes may occur in process variables (water velocity), in variables which express the cross-sectional geometry of the river channel (width, depth, wetter perimeter, hydraulic radius), in measures of sediment accumulations in the channel (dune wavelength, dune height), and in expression for river channel planform (sinuosity, meander arc length). Many of these are inter-related and geometry and planform provide the main variables. To focus upon these morphological consequences of channel adjustment we can visualise a series of four spatial scales at which adjustment may be represented (Gregory, 1976) and these range from the sedimentary bedforms in the channel, through the river channel cross-section, and the river channel planform, to the drainage network. There can be adjustment of size, shape, and composition at any of these scales and examples of each type are illustrated in Table 8.3. It must be remembered that an adjustment of a specific river channel may involve changes indicated in any one of the nine categories (Table 8.3). Thus an increase in peak discharges due to the effects of urbanization could, for example, lead to an increase of

Table 8.2
DEGREES OF FREEDOM OF A RIVER CHANNEL

Adjustment of width	may occur in response to input of water
depth	sediment
slope	bed and bank material
velocity	valley slope
plan shape	

Source: Hey (1974)

To define bankfull hydraulic geometry we need to simultaneously solve 9 process equations in which the dependent variables are:

bankfull hydraulic radius
channel slope
bankfull dune wavelength
bankfull dune height
bankfull wetted perimeter
maximum bankfull depth
channel sinuosity
meander arc length

Source: Hey (1978)

channel cross-section size, an alteration of channel shape, an increase in the size of meanders and metamorphosis of planform along selected reaches from single thread to multithread as sedimentary bars accumulated. Although changes of each of the individual components included (Table 8.3) have been recorded associated with changes in river channel process, there is an outstanding need to understand the exact way in which the several possible adjustments inter-relate and interact during the river metamorphosis. The categories of adjustment proposed (Table 8.3) can be used to elucidate changes of river channel cross-section, of channel planform or pattern, and of drainage networks.

8.3a Adjustments of river channel cross sections

Downstream of dams and reservoirs the magnitude and frequency of peak discharges is often decreased by the effect of water storage, and sediment is trapped in the reservoir. Water released from the dam may scour the channel immediately below the dam and this has to be allowed for when the dam is designed in order to avoid the possibility of erosion undermining the dam (Komura and Simons, 1967). Maximum degradation below the dam can be up to 15cm per year but it may occur along a channel length equal to 69 channel widths (Wolman, 1967a). Adjustment to bed degradation was apparent for nearly 250 km below Elephant Butte dam in the USA (Stabler, 1925). The importance of a

Table 8.3

POTENTIAL RIVER CHANNEL ADJUSTMENTS

Potential adjustments of:	Fluvial landform		
	River channel cross section	River channel pattern	Drainage network
Size	INCREASE OR DECREASE OF RIVER CHANNEL CAPACITY Erosion of bed and banks can produce a larger channel which maintains the same shape Sedimentation can produce a smaller channel which maintains the same shape.	INCREASE OR DECREASE OF SIZE OF PATTERN Increase or decrease of meander wavelength whilst preserving the same planform shape.	INCREASE OR DECREASE OF NETWORK EXTENT AND DENSITY Extension of channels or shrinkage of perennial, intermittent and ephemeral streams.
Shape	ADJUSTMENT OF SHAPE Width/Depth ratio may be increased or decreased	ALTERATION OF SHAPE OF PATTERN A change from regular to irregular meanders	DRAINAGE PATTERN CHANGED IN SHAPE Inclusion of new stream channels after deforestation
Composition	CHANGE IN CHANNEL SEDIMENTS Alteration of grain size of sediments in bed and banks possibly accompanied by development of berms or bars.	PLANFORM METAMORPHOSIS Change from single to multithread channel or converse.	NETWORK COMPOSITION CHANGED The replacement of channels with no definite stream channel (dambos in West Africa) by a clearly defined channel.

General changes are indicated in capitals and examples given in lower case letters.

Source: Gregory (1976)

reduced peak discharge downstream of this dam on the Rio Grande had induced changes in channel size after storage operations were initiated in 1915 (Sonderreger, 1935). A number of studies have now been undertaken to show how channel capacities are smaller than expected downstream of dams, and downstream of Clatworthy reservoir on the Tone, Somerset the channel capacities were shown (Gregory and Park, 1974) to be less than half the expected size downstream of the dam (Fig. 8.3A) and the effect of the reservoir was apparent downstream until the drainage area was 2·5 times that draining to the reservoir. Studies have also investigated the way in which channel size reduction is achieved, and in the Willow Creek Basin of Montana detention reservoirs were shown to be the cause of a reduction in channel width (Frickel, 1972). The regulated discharge downstream of dams can induce changes in vegetation limits and sometimes in the encroachment of riparian vegetation. Downstream of the dams constructed between 1899 and 1943 in the headwaters of the river Derwent in the southern Pennines, a river channel capacity reduced to 40 per cent of its former level occurred as a result of the formation of a bench along certain reaches and this bench has been dated by reference to large-scale maps and plans and by using tree ring dating of trees growing on the bench compared with those on the margin of the flood plain proper (Petts, 1977).

Downstream of urban areas, channel capacities may increase as a result of the increased frequency of flood discharges. In Philadelphia, Pennsylvania, comparison of channel capacities in relation to age and character of urban area indicated (Hammer, 1972) that channels within and downstream of urban areas can be up to four times the size of their rural counterparts. If river discharge is the only variable to change, then the size of stream channels is expectably increased, but sediment availability is greatly increased during building operations but drastically reduced when urbanization is complete. Thus along the Watts Branch, Maryland, it was shown by repeated surveys of river channel cross-sections (Fig. 8.3B) that capacities declined for twelve years during building operations but later began to increase (Leopold, 1973) (Fig. 8.3). If an urban area provides sediment to a major river channel then aggradation may occur and along Dumaresq Creek, New South Wales, channel capacities were shown to be as little as 0·13 (Fig. 8.3C) of the size expected (Gregory, 1977b). An important consideration in the erosion of urban channels is the impact of individual storms. This has been illustrated in the Patuxent Basin, Maryland where catastrophic floods widened the urban channels by 50 per cent more than rural ones, where the size and shape of urban channels change at rates at least three times greater than those in comparable rural areas, where urban channels hold about fifteen times as much sediment as rural channels, and deposition and scour are more severe (Fox, 1976).

Associated with such changes in stream-channel size have been changes in channel shape and below urban areas it has been argued that width may change more easily than depth because of the armouring which may occur on the channel bed. Where discharges are increased the channel may be incised so that the width–depth ratio is decreased. Thus streams in a number of areas of New South Wales may have become incised since a modified runoff regime was occasioned by nineteenth-century settlement and colonization (Dury, 1968), and in Wisconsin, Trimble (1976) has analysed channel change since initial settlement (Fig. 8.3D).

Many of these changes in river-channel geometry have involved alterations of sediment

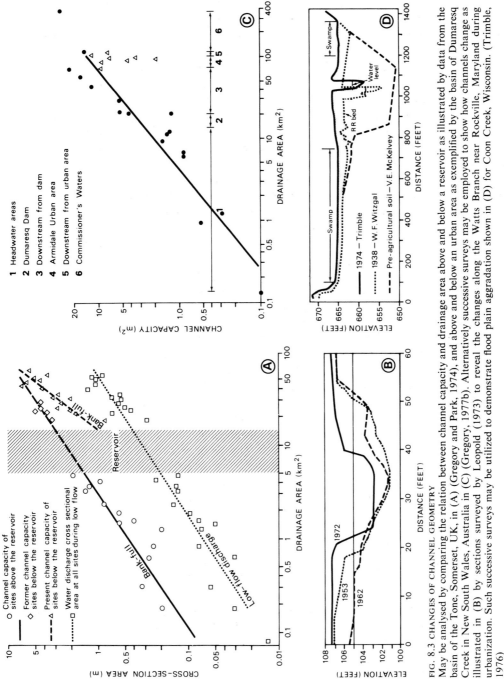

FIG. 8.3 CHANGES OF CHANNEL GEOMETRY

May be analysed by comparing the relation between channel capacity and drainage area above and below a reservoir as illustrated by data from the basin of the Tone, Somerset, UK, in (A) (Gregory and Park, 1974), and above and below an urban area as exemplified by the basin of Dumaresq Creek in New South Wales, Australia in (C) (Gregory, 1977b). Alternatively successive surveys may be employed to show how channels change as illustrated in (B) by sections surveyed by Leopold (1973) to reveal the changes along the Watts Branch near Rockville, Maryland during urbanization. Such successive surveys may be utilized to demonstrate flood plain aggradation shown in (D) for Coon Creek, Wisconsin. (Trimble, 1976)

transport, deposition, and temporary storage, and so the composition of the channel cross-section has changed. This was clearly illustrated in the Meadow Hills area of Denver Colorado where suburban development led first to an expansion of floodplains (Fig. 8.4B) which was later followed by downcutting of streams (Graf, 1975). The inter-relation of changes of capacity, channel shape, and channel composition still requires further elucidation and the channels of a particular area at any one time may be at a particular point in their adjustment process. This is illustrated in the Ruahine Range of New Zealand where forest clearance, for example along the Tamaki river, saw channel width increased from 5 to 10 m in the early 1920s, to 54 m by 1942 and to 60 m by 1976 (Mosley, 1978). In this area the increase of sediment supply rates from the deforested

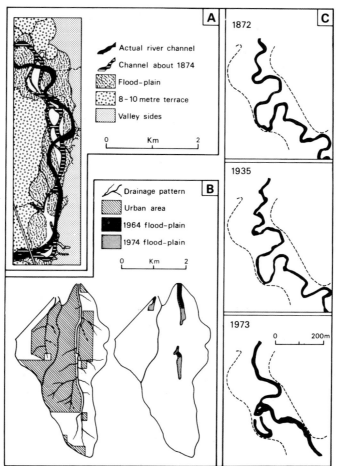

FIG. 8.4 CHANGES OF CHANNEL PATTERN
(A) illustrates the changes documented for a section of the Wisłoka River, Polish Carpathians. Nineteenth-century changes in land use led to the development of a multi-thread channel in the nineteenth century. (Source: Klimek and Starkel, 1974). (B) indicates the extension of flood plain identified by Graf (1975) in the Denver area of Colorado, USA which followed suburban development and the production of large quantities of sediment. Changes in the course of a section of the River Bollin, Cheshire, UK were analysed by Mosley (1975) as illustrated in (C) to reflect changes in land use and flood frequency.

valley throats and alluvial fans at the foot of the range has induced stream channel trenching. Thus the influence of vegetation on the channel banks, the changes in discharge, and the increased sediment supply, all have to be considered in relation to interpretation of channel adjustment.

8.3b Changes of river channel planform

Two kinds of change of pattern may be distinguished (Lewin, 1977): namely *autogenic* which are inherent in the river regime, and *allogenic* which occur in response to changes in the system, including those induced by man. It is often difficult to completely differentiate between these two types, but change in the size of channel pattern is a consequence of increased frequency of flood events. A detailed study of river planforms in Devon was made (Hooke, 1977) by comparing patterns on tithe survey maps with those on several editions of large-scale Ordnance Survey maps. This revealed changes in planform between *c.* 1840 and 1903, including a decrease in mean wavelength along 75 per cent of the sections studied and further changes along 69 per cent of the sections between 1903 and 1958. There was also an increase in sinuosity in this period. Thus the size of the pattern was changing and this was also accompanied by a change in regularity although the influence of human interference is difficult to separate from other variables. The White River, Indiana, has been analysed by comparing reaches of the river from air photographs taken in 1937 and 1966–8 and with a map published in 1880 (Brice, 1974). From this comparison it was concluded (Brice, 1974) that the pattern became less regular and sinuous after 1880 and this may have been the result of accelerated migration following nineteenth-century deforestation of the flood plain. Spectral analysis of the development of the pattern also confirms the decline in regularity (Ferguson, 1977). Examples have been described where a river shape changes following man's activity and along the Dunajee river in the Carpathians of southern Poland comparison of surveys made in 1787, 1880, and 1955 showed how the channel was becoming more sinuous and branching; and this change of shape has been associated with deforestation and with the advent of potato cultivation in the early nineteenth century because this allowed high rates of slope runoff (Klimek and Trafas, 1972). A similar change has been identified along the Wisłoka Valley (Fig. 8.4A). Changes of shape may result from the confinement of river channel patterns by man, and Lewin and Brindle (1977) have shown how the incidence of rail, road, and other engineering works may retard the autogenic migration of meanders downstream and modify meander shapes, and how in other cases the loops become confined and cutoffs occur. A change in shape may often depend for its execution upon the occurrence of flood events which exceed the critical threshold value. Thus the channel of the River Bollin in Cheshire (Fig. 8.4C) was stable between 1872 and 1935, but then rates of channel shifting were increased and seven meander cutoffs contributed to a decline in sinuosity from 2·34 to 1·37 (Mosley, 1975). Although this change in shape was triggered by the large floods of the 1930s it may have been a response to increased runoff following agricultural land drainage and urban development in the area in the twentieth century (Mosley, 1975).

Planform metamorphosis is most dramatic when the channel changes from one type of pattern to another. The South Platte River, USA, has always been cited as a classic example of a braided stream, but whereas the channel was about 800 m wide in 1897 about 90 km above its junction with the North Platte River, it had narrowed to 60 m by 1959 (Schumm, 1977), and the tendency has been for a single thread channel to replace the former braided and less sinuous one. Along this river and along the North Platte and Arkansas rivers there have been flood control works and diversions for irrigation, and the narrowing of the North Platte river is associated with a decrease in the mean annual flood from 370 m^3s^{-1} to 85 m^3s^{-1} (Schumm, 1977).

8.3c   Drainage network adjustments

Changes of the drainage network may accompany changes in channel geometry and channel pattern induced by man. The most dramatic changes are those which arise from extension of the network, and gully development has been well documented (e.g. Cooke and Reeves, 1976) and is discussed in chapter 12. The metamorphosis of channel patterns described in the United States (Schumm, 1977) is often associated with the incision of stream channels and the development of gullies as illustrated by the Strahler model (Fig. 8.2A). Less widespread are the changes which result from contraction but these occur where the water table has been lowered so that perennial or intermittent streams cease to flow. Metamorphosis of the drainage network must be considered in terms of the dynamic elements of the network. The relative significance of perennial, intermittent, and ephemeral streams may change, so that it is not merely a question of considering the extent of the network but it is also desirable to consider the function of the streams composing the network as well.

Alterations of drainage pattern have occurred widely both directly and indirectly. Man's direct influence is shown where drainage ditches are dug so that the pattern of the stream network is effectively changed, and in England and Wales land-drainage channels have an estimated length of 128 000 km (Marshall, Wade and Clare, 1978). In areas recently afforested the pattern may have been amplified by drainage channels, so that in the Coalburn catchment, Northumberland, the stream density was increased by about fifty times (Institute of Hydrology, 1973). Indirect changes of the drainage network arise when a change in land use produces new elements in the pattern and this is illustrated where a land-use change has introduced new rills or streams to complement the existing network.

Such changes may be cases of adjustments in the composition of the drainage network. In the seasonal tropics, there are depressions without a stream channel but along which water flow occurs in the wet season, and these are called dambos in West Africa. If land-use pressure induces changes in runoff rates this may be expressed in the production of a stream channel along such depressions and this may develop to a valley floor gully. An illustration of a change in the composition of the drainage network is provided on the Southern Tablelands of New South Wales. At the time of exploration and initial settlement, many drainage lines contained chains of ponds, but these were often destroyed by channel entrenchment between 1840 and 1950 as tree removal and grazing pressure affected the area (Eyles, 1977). This change involved a sequence of changes from chain of 'scour' ponds, to discontinuous gully, to continuously incised channel, to channel containing 'fixed bar' ponds, to permanently flowing stream (Fig. 8.5B). Only recently have such changes been identified but they may have been more widespread than

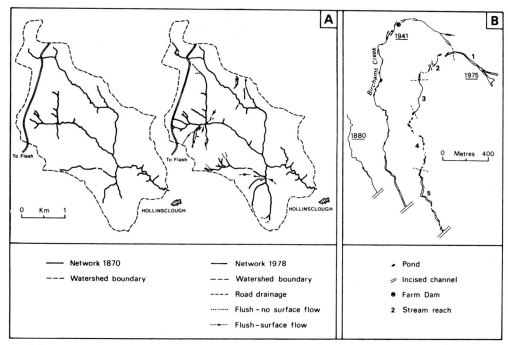

FIG. 8.5 EXAMPLES OF NETWORK CHANGE
(A) illustrates the change in the drainage network of the head of the Dove basin southern Pennines, UK. The 1870 network is based upon the first edition 1:10560 map and the network in 1978 is based upon the 1970 map and field survey in 1978 (see Gregory and Ovenden, 1979). (B) shows the changing channel form of Birchams Creek, a tributary of the Yass river north east of Canberra ACT, Australia. The creek in 1880 is based upon Surveyor Potter; aerial photographs were used for the 1941 pattern and field survey by R. J. Eyles produced the 1975 pattern. (Source: R. J. Eyles, 1977)

previously envisaged. In upland Britain many headwater channels have *juncus*-lined depressions without a stream channel but comparison of nineteenth and twentieth century maps indicates that these depressions without a stream channel may have become less frequent during the last century (Fig. 8.5A). As drainage from farms, tracks and metalled roads has become more prominent, stream channels have developed where none existed previously (Gregory and Ovenden, 1979).

It is now apparent that interpretation of the metamorphosis of fluvial landforms must consider the inter-relationships between channel geometry, channel planform, and drainage network (Table 8.3). This is evident in south-western Wisconsin where many headwater and tributary channels now have relatively wide and shallow cross-sections compared with the pre-settlement channels (Knox, 1977), and also in New South Wales where the drainage system of Dumaresq Creek includes evidence of change from dambo to stream channel, indications of reduced channel capacities consequent upon reservoir construction, and clear signs of metamorphosed channels arising from urban runoff (Gregory, 1977b). Similar inter-relationships (Table 8.4) may have prevailed in other areas, although they are more difficult to decipher if the time scale of man's activities has been greater and the controlling influences themselves more complex.

8.4 RESTRAINING RIVERS

To control river flow and to prevent channel adjustment a variety of engineering methods can be employed (Table 8.4) and these include channel control structures which are designed primarily to prevent erosion of an existing bank, and training structures which are used to guide the flow and/or to promote deposition of sediment in specific areas (Vanoni, 1975). The channel pattern can also be deliberately modified (Table 8.4), as in the case of the Yangtze where one reach of the river near Tung-t'ing Hu has been reduced from 240 to 160 km by artificial cutoffs (Fig. 8.6). In addition to these types of planned project there are also emergency works following catastrophic storms. Thus after Hurricane Camille in 1969 many reaches of stream channels in Virginia were cleared and straightened independently of any major planned programme (Keller, 1976). Other techniques of channel 'improvement' include channel clearing and deliberate alteration of channel shape and size (Table 8.4).

These various techniques of channel improvement, collectively referred to as channelization, have been applied to many locations along world rivers. The original purpose of the channelization of a specific reach was to prevent erosion or flooding at that location but subsequently it has been realised that channelization procedures can have implications beyond the reach of river which is controlled and that integrated approaches to river planning and control should be further developed.

Table 8.4
ENGINEERING METHODS TO RESTRAIN RIVERS

Type	(a) Channel control structures Method
Revetment	utilized primarily for bank protection and usually applied to bank previously sloped or shaped to a designed form.
blanket revetment	constructed of rock, concrete, asphalt, or other materials
pervious revetment	open fence, baskets
solid fence	one or more rows of fencing backed by other materials
groins	short solid structures at right angles to flow
Training structures	employed to guide flow so that effective channel will be scoured and maintained along required course
timber pile dikes	single or multiple rows of piling
jacks	
rock dikes	
rock-filled pile dikes	
Closure of chutes or secondary channels	where flow is precluded in a secondary channel or diverted from a major to a minor channel or a new channel is cut (as in the case of a cutoff)

Type	(b) Channel modification Method
Emergency clearing	following catastrophic storms, bulldozing of channels to remove debris and sediment and to straighten channels
Change of channel geometry	widening, deepening or straightening, often involving removal of trees and bank vegetation
Clearing of channel	removal of debris, trees, and obstructions, leading to increased flow velocities

Source: Vanoni (1975)

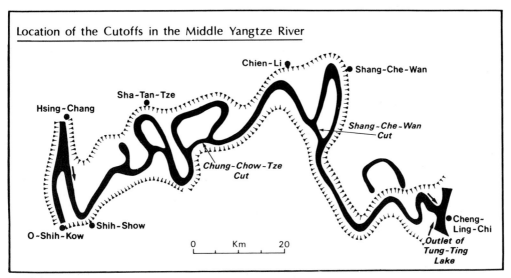

FIG. 8.6 CUTOFFS ALONG THE YANGTZE, CHINA
The two meanders were cut off in 1968 and the effects upon sediment discharge and upon the river channel have been investigated by Pan Ching-shen *et al.,* 1977

8.4a Consequences of channelization

Although the effects of channelization are immediately evident in the river reach which is protected, the ancillary consequences both upstream and downstream take longer to become apparent. Extensive reaches of the East and West Prairie rivers in Alberta, Canada, were channelized and straightened between 1953 and 1971 to provide flood relief. However upstream of the channelized portions, nearly 5 m of degradation occurred between 1964 and 1974, and there has also been aggradation downstream which is reducing the capacity of the channels to cope with floods (Parker and Andres, 1976). A specific instance of adjustment consequent upon channel relocation was provided along the Peabody River in New Hampshire where the shortening of the river course gave a slope of 15·2 m km^{-1} replacing the former slope of 9·6 m km^{-1} and the river had reduced this to 13·3 m km^{-1} after seven years by degradation at the upstream end and aggradation downstream (Yearke, 1971).

A large literature has now developed about the implications of channelization but the main ones have been identified as downstream effects, damage to channel and floodplain, damage to fish and wildlife, and aesthetic degradation (Keller, 1976). One type of downstream effect is indicated by the examples from Alberta, but this will not always apply because flooding and sedimentation could be reduced downstream. Damage to channel and floodplain can occur because significant bank erosion may be induced by channelization procedures and this has been illustrated for the headwaters of the Blackwater River in Missouri where a new channel was dredged in 1910: the channel increased from a cross sectional area of 19 m^2 to between 160 and 484 m^2, channel widths increased from about 15 m to 60 m, and bridges had to be rebuilt (Emerson, 1971). Many changes have been inspired by changes of slope, because if these are too severe then the river proceeds to reduce slope by upstream degradation and sedimentation downstream. Channelization may remove the pools and destroy the vegetation and also increase the velocities such that habitats for fish and wildlife are dramatically changed. The general loss of the picturesque and the varied features of a river consequent upon improvement schemes has been termed aesthetic degradation (Keller, 1976).

8.4b   Integrated approaches

Channelization was first used for specific-problem reaches without concern for the effects on the channel system as a whole. At a time when concern for environmental quality has become very apparent it is evident that changes in a river system must be designed against the background of the character of the river system and its basin as a functioning unit. This necessitates a greater physical understanding of the present mechanics of rivers and of the way in which they react to change; use of an approach which is based upon integrated consideration of the river, channel, floodplain, and basin as a system; and an assessment of water resource projects in relation to environmental values (Keller, 1976). It is therefore possible to envisage a number of ways whereby the channelization procedures can be less obtrusive in the landscape (Table 8.5) and designed methods should be based upon a clear understanding of river behaviour by river regulation with the aid of nature (Winkley, 1972). It is desirable to apply an appropriate classification of the zones of a river system, to know the interactions of form and process in each zone, and to plan the future developments of the river in its basin against this zonal pattern. Thus Palmer (1976) advocated reference to geohydraulic river environment zones which are of four types along the rivers of Washington and Oregon: namely the boulder zone (headwater) the floodway zone, pastoral zone, and estuarine zone.

8.5   PROSPECTS FOR RIVERS AND MAN

A recent article by L. B. Leopold (1977) who has studied rivers for many years was entitled 'a reverence for rivers'. Whereas rivers were originally conceived as an element of the landscape that could be altered at specific sites as required, research has now shown how the river functions as a system, and how it may react not only in the area affected directly by man but elsewhere as well. Studies have been made of the quantity and quality of the water flowing along the channel, of the quantitative aspects of the river channels and their changes, and it now remains to look more closely at the quality of riverscape. We can only maintain this quality, or enhance it, by working with the river and not against it and hence we must continue to have a reverence for rivers!

Table 8.5
SOME METHODS TO RETRAIN RIVERS

Use revetments or training structures arranged to provide a variety of flow conditions
e.g. artificial pools and riffles

Channelize one bank of stream only

Construct diversion channel to allow flood flows to bypass natural channel which is maintained

---

See Table 8.4

REFERENCES

BEAUMONT, P., 1978, 'Man's impact on river systems; a world wide view.' *Area* 10, pp. 38–41.

BECKINSALE, R. P., 1972, 'The effect upon river channels of sudden changes in load.' *Acta Geographica Debrecina*, 10, pp. 181–6.

BLENCH, T., 1972, 'Morphometric changes', *River Ecology and Man*, ed. R. T. OGLESBY, C. A. CARLSON and J. A. MCCANN (Academic Press, New York), pp. 287–308.

BRICE, J. C., 1974, 'Meander pattern of the White River in Indiana – an analysis', *Fluvial Geomorphology*, ed. M. E. MORISAWA (State University of New York, Binghamton), pp. 178–200.

BUTZER, K. W., 1971, 'Fine alluvial fills in the Orange and Vaal basins of South Africa.' *Proc. Assoc. Amer. Geog.*, 3, pp. 41–8.

CHING-SHEN, PAN, SHAO-CHUAN, SHI and WEN-CHUNG, TUAM, 1977, *A Study of the Channel Development after the Completion of Artificial Cutoffs of the Middle Yangtze* (Peking).

CHOU, TSUNG-LIEN, 1976, 'The Yellow River: its unique features and serious problems.' *Rivers '76,* Amer. Soc. Civil Eng., pp. 507–26.

COOKE, R. U. and REEVES, R. W., 1976, *Arroyos and Environmental Change in the American South-West* (Clarendon Press, Oxford).

CRAMPTON, C. B., 1969, 'The chronology of certain terraced river deposits in the south-east Wales area.' *Zeitschrift für Geomorphologie*, 13, pp. 245–59.

DOLAN, R., HOWARD A. and GALLENSON, A., 1974, 'Man's impact on the Colorado River in the Grand Canyon.' *American Scientist*, 62, pp. 392–401.

DURY, G. H., 1968, Footnote in WOODYER, K. D., 1968, 'Bankfull frequency in rivers.' *Jnl. Hyd.,* 6, pp. 114–42.

EMERSON, J. W., 1971, 'Channelization: a case study.' *Science*, 1973, pp. 325–6.

EYLES, R. J., 1977, 'Changes in drainage networks since 1820. Southern Tablelands, NSW.' *Australian Geographer*, 13, pp. 377–86.

1977, 'Birchams Creek: the transition from a chain of ponds to a gully.' *Austral. Geog. Studies*, 15, pp. 146–57.

FERGUSON, R., 1977, 'Meander migration: equilibrium and change', *River Channel Changes*, ed. K. J. GREGORY (Wiley, Chichester), pp. 235–48.

FOX, H. L., 1976, 'The urbanizing river; a case study in the Maryland Piedmont', *Geomorphology and Engineering*, ed. D. R. COATES (State University of New York, Binghamton), pp. 245–71.

FRICKEL, D. G., 1972, 'Hydrology and effects of conservation structures, Willow Creek Basin, Valley County, Montana 1954–68.' *US Geol. Surv. Water Supply Pap.*, 1532–G.

GRAF, W. L., 1975, 'The impact of suburbanization on fluvial geomorphology.' *Water Resources Res.*, 11, pp. 690–2.

1977, 'The rate law in fluvial geomorphology.' *Amer. Jnl. Sci.*, 277, pp. 178–91.

GREGORY, K. J., 1976, 'Changing drainage basins.' *Geog. Jnl.*, 142, pp. 237–47.

1977a 'The context of river channel changes', *River Channel Changes*, ed. K. J. GREGORY (Wiley, Chichester), pp. 1–12.

1977b, 'Channel and network metamorphosis in northern New South Wales', *River Channel Changes*, ed. K. J. GREGORY (Wiley, Chichester), pp. 389–410.

GREGORY, K. J. and OVENDEN, J. C., 1979, 'The permanence of stream networks in Britain', *Earth Surface Processes* (J. Wiley, Chichester), in press.

GREGORY, K. J. and PARK, C. C., 1974, 'Adjustment of river channel capacity downstream from a reservoir.' *Water Resources Res.*, 10, pp. 870–3.

HAMMAD, H. Y., 1972, 'Riverbed degradation after closure of dams.' *Proc. Amer. Soc. Civ. Engrs.*, *Jnl. Hyd. Div.*, 98, pp. 591–607.

HAMMER, T. R., 1972 'Stream channel enlargement due to urbanization.' *Water Resources Res.*, 8, pp. 1530–40.

HEY, R. D., 1974, 'Prediction and effects of flooding in alluvial systems', *Prediction of Geological Hazards*, ed. B. M. FUNNELL Geol. Soc. Misc. Pap. 3, pp. 42–56.

1978, 'Determinate hydraulic geometry of river channels.' *Proc. Amer. Soc. Civ. Engrs., Jnl. Hyd. Div.,* 104, pp. 869–85.

HOOKE, J. M., 1977, 'The distribution and nature of changes in river channel patterns. The example of Devon', *River Channel Changes,* ed. K. J. GREGORY (Wiley, Chichester), pp. 265–80.

INSTITUTE OF HYDROLOGY, 1973, *Research 1972–3* (Natural Environment Research Council).

KELLER, E. A., 1976, 'Channelization: environmental, geomorphic and engineering aspects', *Geomorphology and Engineering,* ed. D. R. COATES (State University of New York, Binghamton), pp. 115–40.

KLIMEK, K. and STARKEL, L., 1974, 'History and actual tendency of floodplain development at the border of the Polish Carpathians.' *Abh. d. Akademie der Wiss in Gottingen, Math-Phys. Klass,* 3, 29, pp. 185–96.

KLIMEK, K. and TRAFAS, K., 1972, 'Young Holocene changes in the course of the Durajec River in the Beskia Sadecki Mts (Western Carpathians).' *Studia Geomorphologica Carpatho-Balcanica,* 6, pp. 85–92.

KNOX, J. C., 1977, 'Human impacts on Wisconsin stream channels.' *Ann. Assoc. Amer. Geog.,* 67, pp. 323–42.

KOMURA, S. and SIMONS, D. B., 1967, 'River bed degradation below dams.' *Proc. Amer. Soc. Civ. Engrs, Jnl. Hyd. Div.,* 93, pp. 1–14.

LANE, E. W., 1955, 'The importance of fluvial morphology in hydraulic engineering.' *Proc. Amer. Soc. Civ. Engrs.,* 81, pp. 1–17.

LEOPOLD, L. B., 1973, 'River channel change with time: an example.' *Bull. Geol. Soc. of America,* 84, pp. 1845–60.

1977, 'A reverence for rivers.' *Geology,* pp. 429–30.

LEWIN, J., 1977, 'Channel pattern changes', *River Channel Changes,* ed. K. J. GREGORY (Wiley, Chichester), pp. 167–84.

LEWIN, J. and BRINDLE, B. J., 1977, 'Confined meanders', *River Channel Changes,* ed. K. J. GREGORY (Wiley, Chichester), pp. 221–34.

LEWIN, J., DAVIES B. E. and WOLFENDEN, P. J., 1977, 'Interactions between channel change and historic mining sediments', *River Channel Changes,* ed. K. J. GREGORY (Wiley, Chichester), pp. 353–68.

MARSHALL, E. J. P., WADE, P. M. and CLARE, P., 1978, 'Land drainage channels in England and Wales.' *Geog. Jnl.,* 144, pp. 254–63.

MOORE, C. M., 1969, 'Effects of small structures on peak flow', *Effect of Watershed Changes on Stream flow,* ed. C. M. MOORE and C. W. MORGAN (Univ. Texas Press, Austria), pp. 101–17.

MOSLEY, M. P., 1975, 'Channel changes on the River Bollin, Cheshire 1872–1973.' *East Midland Geographer,* 6, pp. 185–99.

1978, 'Erosion in the south-eastern Ruahine Range; its implications for downstream river control.' *NZ Jnl. Forestry,* 23, pp. 21–48.

MROWKA, J. P., 1974, 'Man's impact on stream regimen and quality', *Perspectives in Environment,* ed. I. R. MANNERS and M. W. MIKESELL, Assoc. Amer. Geog., Publ., 13, pp. 79–104.

O'LOUGHLIN, C. L., 1970, 'Streambed investigations in a small mountain catchment.' *NZ Jnl. Geol. Geophys.,* 12, pp. 684–706.

PAINTER, R. B. *et al.,* 1974, 'The effect of afforestation on erosion processes and sediment yield.' *Proc. Symp. Effects of Man on the Interface of the Hydrological Cycle with the Physical Environment,* Int. Assoc. Sci. Hyd. Publ., 117, pp. 62–7.

PALMER, L., 1976, 'River management criteria for Oregon and Washington.' *Geomorphology and Engineering,* ed. D. R. COATES (State University of New York, Binghamton), pp. 329–46.

PARKER, G. and ANDRES, D., 1976, 'Detrimental effects of river channelization.' *Rivers '76,* Amer. Soc. Civ. Engrs., pp. 1248–66.

PEMBERTON, E. L., 1976, 'Channel changes in the Colorado river below Glen Canyon dam.' *Proc. Third Inter Agency Sedimentation Conf.* pp. 5–61 to 5–73.

PETTS, G. E., 1977, 'Channel response to flow regulation: the case of the River Derwent, Derbyshire', *River Channel Changes,* ed. K. J. GREGORY (Wiley, Chichester), pp. 145–64.

RICHARDSON, E. V. and CHRISTIAN, H., 1976, 'Channel improvements on the Missouri River.' *Proc. Third Fed. Inter Agency Sedimentation Conf.* pp. 5–113 to 5–124.

RICHARDSON, E. V. and SIMONS, D. B., 1976, 'River response to development.' *Rivers '76,* Amer. Soc. Civ. Engrs., pp. 1285–1300.

RUTTER, E. J. and ENGSTROM, L. R., 1964, 'Reservoir regulation', *Handbook of Applied Hydrology,* ed. V. T. CHOW (McGraw Hill, New York), Section 25.

SCHUMM, S. A., 1969, 'River metamorphosis.' *Proc. Amer. Soc. Civ. Engrs., Jnl. Hyd. Div.,* 95, pp. 251–73.

1971, 'Channel adjustment and river metamorphosis', *River Mechanics,* I, ed. H. W. SHEN (Water Resources Publications, Fort Collins), pp. 5–1 to 5–22.

1977, *The Fluvial System* (Wiley, New York).

SHERLOCK, R. L., 1922, *Man as a Geological Agent* (Witherby, London).

SONDERREGER, A. L., 1935, 'Modifying the physiographic balance by conservation measures.' *Trans. Amer. Soc. Civ. Engrs.,* Paper No 1897, pp. 284–304.

STABLER, H., 1925, 'Does desilting affect cutting power of streams?' *Eng. News Records,* 95, p. 960.

STEVENS, M. A., SIMONS, D. B., and SCHUMM, S. A., 1975, 'Man-induced changes of Middle Mississippi River.' *Proc. Amer. Soc. Engrs., Jnl. Waterways Harbors, Coast. Eng. Div.,* 101, pp. 119–33.

STRAHLER, A. N., 1956, 'The nature of induced erosion and aggradation', *Man's Role in Changing the Face of the Earth,* ed. W. L. THOMAS (Univ. of Chicago Press), pp. 621–38.

TODD, D. K., 1970, *The Water Encyclopedia* (Water Information Center, New York).

TRIMBLE, S. W., 1970, 'The Alcovey River swamps. The result of culturally accelerated sedimentation.' *Bull. Georgia Academy of Sci.,* 28, pp. 131–41.

1976, 'Modern stream and valley sedimentation in the Driftless area, Wisconsin, USA' *International Geography '76,* 1, pp. 228–31.

VANONI, V. A. (ed), 1975, *Sedimentation Engineering* (Amer. Soc. Civ. Engrs Task Committee Sedimentation Committee of Hydraulics Division), pp. 531–46.

VITA-FINZI, C., 1969, *The Mediterranean Valleys* (CUP, Cambridge).

WINKLEY, B. R., 1972, 'River regulation with the aid of Nature.' *Int. Comm. Irrigation and Drainage,* Eighth Congress, pp. 433–57.

WOLMAN, M. G., 1967a, 'Two problems involving river channel changes and background observations.' *Quantitature Geography, Part II, North Western studies in Geography,* 14, pp. 67–107.

1967b, 'A cycle of sedimentation and erosion in urban river channels.' *Geog. Annaler,* 49a, pp. 385–95.

YEARKE, L. W., 1971, 'River channel erosion due to channel relocation.' *Civil Engineering,* August 1971, pp. 39–40.

9

Permafrost and Ground Ice

HUGH M. FRENCH
University of Ottawa

9.1 INTRODUCTION TO PERMAFROST DISTRIBUTION AND PROBLEMS

Permafrost, or perennially frozen ground, influences virtually all aspects of man's activities in the regions which it underlies. In North America, the importance of permafrost was realised only recently, when attention focused on Alaska and other northern regions during and immediately after the Second World War. The building of the Alaskan Highway in 1942 from Dawson City to Fairbanks and the construction of the Canol pipeline from Norman Wells to Alaska in 1943 were major engineering undertakings which highlighted the inadequacies of traditional methods of construction. In addition, the difficulties of water and sewage provision, and the limitations placed on agriculture and mining by permafrost meant that large-scale permanent settlement of Arctic North America was unrealistic until permafrost problems were understood. Referring to Alaska, S. W. Muller of the US Army Corps of Engineers, wrote in 1945 'The destructive action of permafrost phenomena has materially impeded the colonisation and development of extensive and potentially rich areas in the north. Roads, railways, bridges, houses and factories have suffered deformation, at times beyond repair, because the condition of permafrost ground was not examined beforehand, and because the behaviour of frozen ground was little, if at all, understood' (Muller, 1945, pp. 1–2).

The Soviet Union, by virtue of its longer history of settlement in permafrost regions, was aware of these problems earlier than other countries and by the late 1920s had established a Permafrost Institute of the Siberian Academy of Sciences at Yakutsk in Eastern Siberia. This institute now employs several hundred permafrost scientists. The nearest equilvalent in North America is the Cold Regions Research and Engineering Laboratory (CRREL) of the US Army, located at Hanover, New Hampshire.

Interest in permafrost regions, especially in the United States, Canada, and the Soviet

Union, has expanded rapidly in the last thirty years. A recent stimulus has been the search for hydrocarbons and other natural resources, particularly in northern Alaska and the Canadian Arctic. Today, a range of government, academic, and private organisations are involved in permafrost related activities in both North America and the Soviet Union. To date, three international permafrost conferences have been held: the first in 1963 in the United States, the second in 1973 at Yakutsk in the Soviet Union, and the third in 1978 at Edmonton in Canada. The proceedings of these meetings constitute some of the most comprehensive summaries of man's activities in permafrost regions presently available.

Permafrost underlies approximately one-fifth of the land surface of the earth. In the northern hemisphere, large areas of permafrost exist in the Soviet Union, Canada, and Alaska (Fig. 9.1). Less important in spatial extent are permafrost occurrences in

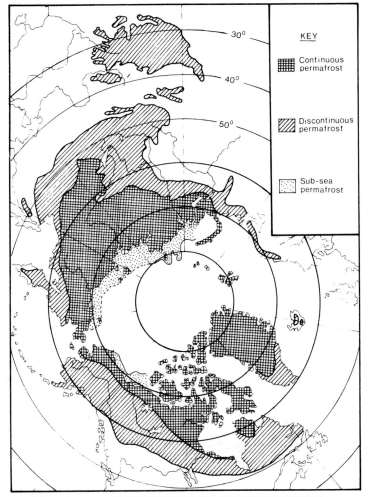

FIG. 9.1 DISTRIBUTION OF PERMAFROST IN THE NORTHERN HEMISPHERE
(Compiled by T. L. Péwé, 1978, from various sources)

Greenland, Spitsbergen and northern Scandinavia. Offshore, or sea-bottom permafrost is also known to exist, particularly in the Beaufort Sea of the Western Arctic and in the Laptev and East Siberian Seas of the Soviet Union. Terrestrial permafrost occurs not only in the tundra and polar desert environments north of the treeline, but also in extensive areas of the boreal forest and forest-tundra environments. In addition, permafrost is present at high elevations in middle latitudes such as the Rocky Mountains of North America and the interior plateaus of central Asia, particularly in Mongolia and Northern China. An unofficial estimate is that approximately 20 per cent of China is underlain by permafrost of one sort or another.

We may classify permafrost as being either continuous or discontinuous in nature. In the continuous permafrost zone, permafrost is present at all localities except for localized thawed zones, or taliks, existing beneath lakes, river channels, and other large water bodies which do not freeze to their bottoms in winter. In the discontinuous permafrost zone, areas of frozen ground are separated by areas of unfrozen ground. At the southern limit of this zone permafrost becomes restricted to isolated 'islands' occurring beneath peaty organic sediments. Permafrost may vary in thickness from a few centimetres to several hundreds of metres. In parts of Siberia and interior Alaska, permafrost has existed for several hundred thousand years; in other areas, such as the modern Mackenzie Delta, permafrost is young and currently forming under the existing cold climate.

Although permafrost is a temperature phenomenon, problems posed by man's activities in permafrost regions usually arise from related characteristics. These may be summarised under four main categories.

First, ground ice is a major component of permafrost, particularly in unconsolidated sediments. Frequently, the amount of ice held within the ground in a frozen state exceeds the natural water content of that sediment in its thawed state. If the permafrost thaws therefore, subsidence of the ground equal in volume to the amount of water released from the soil may result. Thaw consolidation may also occur as the thawed sediments compact and settle under their own weight. In addition, high pore water pressures generated in the process may favour soil instability and mass movement. These various processes associated with permafrost degradation are generally termed thermokarst (e.g. French, 1976, pp. 105–33).

Secondly, the freezing of water in the seasonally frozen zone, commonly termed the active layer, at the beginning of winter each year results in ice lensing and segregation. The volume expansion associated with this phase change from water to ice is approximately 9 per cent. As a result there is an upward expansion of the ground surface each winter, a process termed 'frost heave' (e.g. Washburn, 1973, pp. 65–80). The magnitude of heave varies according to the amount of moisture present in the active layer. Poorly drained silty soils usually possess some of the highest ice or water contents and are termed 'frost susceptible' by engineers.

Thirdly, the physical properties of frozen ground, in which the soil particles are cemented together by pore ice, may be considerably greater than the same material in its unfrozen state (e.g. Tsytovich, 1975). In unconsolidated and soft sediments there is often a significant loss of bearing strength upon thawing.

Fourthly, the hydrologic and groundwater characteristics of permafrost terrain are different to those of non-permafrost terrain (Hopkins, et. al., 1955; Williams and van

Everdingen, 1973). For example, the presence of both perennially and seasonally frozen ground prevents the infiltration of water into the ground or, at best, confines it to the active layer. At the same time, subsurface flow is restricted to unfrozen zones or taliks. A high degree of mineralization in subsurface permafrost waters is often typical, caused by the restricted circulation imposed by the permafrost and the concentration of dissolved solids in the taliks. Thus, frozen ground eliminates many shallow depth aquifers, reduces the volume of unconsolidated deposits or bedrock in which water may be stored, influences the quality of groundwater supply, and necessitates that wells be drilled deeper than in non-permafrost regions.

9.2 GROUND ICE AND THERMOKARST

From the geotechnical and engineering viewpoint, the presence of ground ice is the problem most unique to permafrost regions and central to both thermokarst and frost-heave processes. As a simplification, these problems vary directly with lithology, being most serious in silty unconsolidated sediments and negligible in hard consolidated rock.

Although a variety of ground ice types exist (Mackay, 1972), there are three types which are important in terms of thermokarst, principally because of their ubiquitous nature. The most widespread is pore ice. This is the bonding cement that holds frozen soil grains together. It exists, in varying amounts, in virtually all rock types. Secondly, there is segregated ice which forms lenses ranging from a few centimetres thick to massive ice bodies several metres thick. Thirdly, there is wedge ice which forms from surface water penetrating thermal contraction cracks which develop during winter under the intense cold. Clearly, the total amount of ground ice present varies from locality to locality. In parts of the Western Arctic coastal plain of Alaska and Canada, ice wedges and massive ice bodies assume large dimensions. For example, an estimate from the Point Barrow area in northern Alaska is that wedge ice alone may constitute over 50 per cent of the upper 1–2 m of permafrost (Brown, 1966).

Quantitative estimates of ground ice can be made by either weight or volume. Low ice-content soils are generally regarded as those having ice contents by weight of less than 40–50 per cent. Soils with high ice contents have values which commonly range from between 50 and 150 per cent. From a geomorphic viewpoint, the volume of ice contained by the sediment is important. 'Excess ice' refers to the volume of supernatant water present if the sediment is allowed to thaw. Expressed as a percentage of the total volume of sediment, excess ice values indicate the potential morphological modification or volumetric ground loss consequent upon thawing.

Within this context, thermokarst develops as the result of the disruption of the thermal equilibrium of the permafrost and an increase in the depth of the active layer. This can be illustrated with a simple example (Fig. 9.2). Consider a well vegetated tundra soil with an active layer of 45 cm underlain by supersaturated permafrost which yields upon thawing, on a volumetric basis, 50 per cent excess ice (water) and 50 per cent saturated soil. If the top 15 cm of soil is removed, the thermal insulating role played by the organic mat disappears. Under the bare-ground conditions that result, the depth of seasonal thaw might increase to 60 cm. Since only 30 cm of the original active layer remains, a further

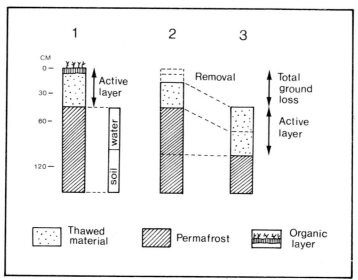

FIG. 9.2 DIAGRAM ILLUSTRATING HOW TERRAIN DISTURBANCE OF AN ICE-RICH TUNDRA SOIL CAN LEAD TO THERMOKARST SUBSIDENCE
(Source: Mackay, 1970)

60 cm of the permafrost must thaw in order to increase the active layer thickness to 60 cm since 30 cm of supernatant water will be released. Thus, in addition to the original 15 cm of material loss from the surface, the terrain subsides a further 30 cm before a new thermal equilibrium is reached.

If we ignore large-scale climatic fluctuations, an infinite number of local geomorphic and vegetation changes may affect thermokarst (Fig. 9.3). These may be either natural or man-induced. For example, natural causes of thermokarst in areas north of the treeline include localised slumping and slope failure, river-bank undercutting, and coastal retreat. In forested permafrost regions, fire is an additional factor since by causing rapid changes in vegetation cover the ground thermal regime can be altered substantially. For example, in a part of the Siberian taiga, the active layer increased from 40 to 85 cm in twelve years following a fire in 1953 (Czudek and Demek, 1970), and at Inuvik, in the Northwest Territories of Canada, where a forest fire occurred in 1968, a 40 per cent increase in the active layer depth in the burned over area was recorded in a four year period (Heginbottom, 1973). Moreover, in an attempt to contain the fire at Inuvik, a number of fire-breaks were bulldozed. In these areas not only were the trees removed but the tops of the underlying frost mounds truncated and the surface organic layer of moss and lichen destroyed. In these areas, thaw depths increased by more than 100 per cent.

Once initiated, thermokarst processes are difficult to arrest. Erosion may become concentrated along ice wedges and continued activity may lead to striking badland topography. In areas where massive icy sediments exist, shallow semicircular slumps may form, the headwalls of which may retreat as much as 8–10 m during a single summer. The cause of these slumps is usually some trigger mechanism associated with the exposure of the permafrost, such as stream undercutting or local slope instability.

One type of mass wasting process favoured in certain permafrost regions is the

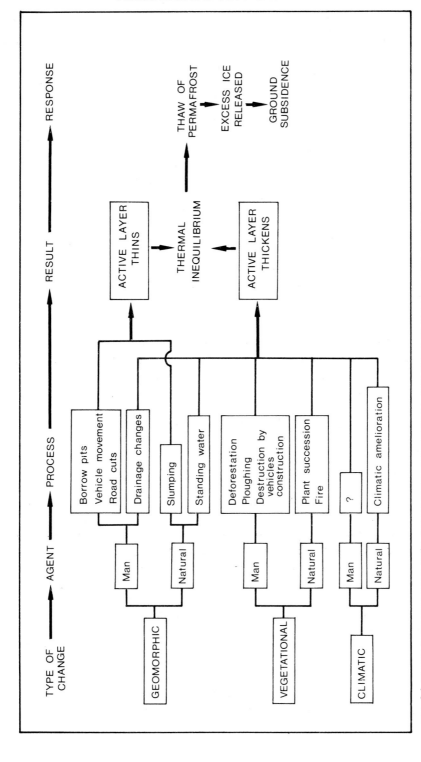

FIG. 9.3 DIAGRAM ILLUSTRATING HOW GEOMORPHIC, VEGETATIONAL AND CLIMATIC CHANGES MAY LEAD TO
PERMAFROST DEGRADATION
(Modified from Mackay, 1970, and French, 1976, p.107)

occurrence of various low-angled failures which are termed 'skin flows' (McRoberts and Morgenstern, 1974). These are confined to the active layer and are induced by high pore water pressures which develop following heavy summer rain. The permafrost table acts as a lubricated slip plane. In instances where the natural water content is close to the liquid limits of the material, multiple shallow mudflows may occur which extend downslope for several hundreds of metres. Particularly large and frequently occurring skin flows are characteristic of the ice-rich shale known as the Christopher Shale Formation which outcrops widely on the Sabine Peninsula of Eastern Melville in the Canadian Arctic (Barnett et. al., 1977). This is an area where sizeable gas deposits have been discovered. Often failure takes place on very low-angled slopes, sometimes less than three degrees. The apparent stability of the tundra surface prior to failure makes the prediction of such events virtually impossible. Since such failures could easily rupture a pipeline traversing such terrain, their occurrence together with the potential for thermokarst makes this terrain exceedingly difficult for construction and other activities.

For reasons such as these, permafrost terrain is often regarded as being highly sensitive to disturbance by human activity, and as presenting unusual engineering and geotechnical problems.

9.3   TERRAIN DISTURBANCES

Our understanding of the consequences of terrain disturbance in permafrost regions has increased significantly in recent years. In nearly all instances, the necessity of maintaining the thermal regime of the permafrost has been recognised and measures to prevent thermal change widely adopted. At the same time, the documentation of case histories of disturbance enables the nature and speed of man-induced thermokarst to be determined and the time period necessary for stabilization to be assessed.

The Soviet Union has by far the greatest experience in this respect. As early as 1925, for example, controlled experiments were being undertaken to determine the effects of vegetation changes on the underlying permafrost, brought about by either deforestation or by ploughing. For example, P. I. Koloskov reported how, in the Yenesei region of Siberia, July soil temperatures at a depth of 40 cm increased by 14°C in a semi-bog soil following deforestation and ploughing (Tyrtikov, 1964). In Alaska, similar experimental studies have been undertaken. One of the earliest was in the Fairbanks region and involved the cutting and/or stripping of the surface vegetation by the US Army Corps of Engineers in 1946. In the stripped area, the active layer increased from 1 m to more than 3 m in thickness over a ten year period. Subsequently, numerous other studies have emphasised the thermal role played by the surface organic layer and/or forest cover (e.g. Babb and Bliss, 1974; Brown et al., 1969; Haugen and Brown, 1970; Kallio and Reiger, 1969). It must be emphasised that even very small disturbances to the surface may be sufficient to induce thermokarst activity. Mackay (1970) for example, describes how an Eskimo dog was tied to a stake with a 1·5 m long chain. In the ten days of tether, the animal trampled and destroyed the tundra vegetation of that area. Within two years, the site had subsided like a pie dish by a depth of 18–23 cm and the active layer thickness had increased by more than 10 cm within the depression.

Without doubt, the most common cause of man-induced thermokarst on a large scale is the clearance of the surface vegetation for agricultural or construction purposes. If this occurs in an area underlain by a polygonal network of ice wedges a distinctive hummocky microrelief forms. This results from the general subsidence of the ground combined with preferential melt along the wedges. A classic example has been described from the Fairbanks region of Alaska where extensive areas were cleared for agriculture in the 1920s. The following thirty years saw the formation of mounds and depressions in the fields, the mounds varying from 3 to 15 m in diameter and up to 2·4 m in height (Rockie, 1942). Subsequently, these features were termed thermokarst mounds (Péwé, 1954). In the Soviet Union, similar undulating surfaces are common in clearings adjacent to small lumber camps in the taiga. In places, shallow troughs, 0·5–1·0 m deep, form hummocky terrain termed 'baydjarakhs' or 'grave-yard mounds' (Soloviev, 1973). In other areas underlain by extremely ice-rich sediments, widespread subsidence may lead to the formation of alas depressions. Adjacent to the village of Maya in central Yakutia for example, a depression 5–8 m deep and between 200 and 300 m in diameter had formed in historic times following deforestation and the beginning of the agricultural settlement (French, 1976, p. 114).

These examples illustrate the extreme sensitivity of permafrost terrain to man-induced surface modifications. We can identify at least three controls over such permafrost degradation. These are first, the ice content of the underlying permafrost and in particular, the presence or absence of excess ice; secondly, the thickness and insulating qualities of the surface vegetation; and thirdly, the duration and warmth of the summer thaw period.

Several case histories illustrate the nature and rapidity of man-induced thermokarst. In all cases, the initial disturbance is associated with borrow pits, where material has been removed for road, airstrip, or other construction purposes. Terrain disturbance at the Deminex Orksut 1–44 wellsite in central Banks Island, in the Western Canadian Arctic, has been monitored since initiation in 1973 (French, 1978a). In August of that year, following completion of the well, the site was abandoned and restoration undertaken. In an attempt to infill the kitchen sump, material was removed from an adjacent gravelly ridge and pushed into the depression. In all, an area of approximately 3000 m² was disturbed (Fig. 9.4A). Two years later a crude polygonal system of gullies had developed at the site, reflecting the preferential melt along underlying ice wedges (Fig. 9.4B). After two further years, in August 1977, typical thermokarst topography had formed, consisting of unstable hummocks and mounds interspersed with standing water bodies (Fig. 9.4C).

The experience at the Orksut 1–44 wellsite can be compared to the man-induced thermokarst which formed adjacent to the airstrip at Sachs Harbour, on southern Banks Island (French, 1975). This airstrip was constructed during summer in the years between 1959 and 1962. In order to grade the proposed strip, thawed material was removed from adjacent terrain and transported, via access ramps, to the strip. In all, a total of 50 000 m² was disturbed and as much as 2·0 m of material removed in places. Subsequent surficial drilling revealed the underlying sediments to be ice-rich sand and gravel with excess ice values of 25–30 per cent and natural water (ice) contents of between 50 and 150 per cent. When first examined in 1972 the borrow pits portrayed actively subsiding thermokarst mound topography. When examined recently in 1978, the mounds were not so sharp and plants were beginning to invade the disturbed terrain.

FIG. 9.4 TERRAIN DISTURBANCES ADJACENT TO EXPLORATORY WELLSITES, WESTERN ARCTIC
(A) (top left) Air view of the Deminex Orksut 1–44 wellsite in central Banks Island, (72°23′N; 122°42′W),
Western Arctic. In an attempt to infill the sump at the living quarters, material was removed from an adjacent
gravelly ridge. (Photograph taken August 1973) (B) (top right) Oblique air view of the Orksut 1–44 wellsite in
July 1975. A system of gullies had developed reflecting preferential melt-out of underlying ice wedges. (C)
(bottom left) Ground view of disturbed terrain at Orksut 1–44 wellsite in August 1977. An unstable topography
consisting of irregular hummocks, depressions and standing water bodies had formed. (D) (bottom right) Gully
erosion which has developed along an old vehicle track made in the summer of 1970 near the site of the Drake
Point blow-out (76°26′N; 108°55′W). The terrain is underlain by ice-rich shales of the Christopher Formation.
(Photograph taken August 1976)

These studies provide some insight into the speed at which man-induced thermokarst develops and the time period over which such terrain remains unstable. They suggest that thermokarst processes are rapid and that the typical hummocky relief forms through preferential subsidence along ice wedges. Stabilization only begins 10–15 years after the initial disturbance and probably is not complete until 30 years or more have passed.

In environments of greater summer thaw than the Western Arctic, an essentially similar sequence of thermokarst activity takes place, except that the amplitude of the thermokarst mounds is greater, probably reflecting the greater thaw depths. For example, an area of disturbed terrain adjacent to the Maya–Abalakh road in Eastern Siberia was examined by the writer in July 1973 (French, 1975, pp. 141–3). Material had been removed to provide aggregate for the road. At one locality where construction had taken place three years previously, mounds similar in size to those at Sachs Harbour had formed. In a second area where disturbance had occurred in 1966, the mounds were much larger, with relative relief exceeding 3 m, and were interspersed with deep standing water bodies reflecting the active melt of the ice-rich permafrost beneath.

A second major cause of man-induced thermokarst relates to the movement of vehicles over permafrost terrain. If this occurs in summer when the surface has thawed and is soft and wet, surface vegetation can be destroyed and deep trenching and rutting can occur. Probably the worst examples of this sort of activity exist on the Alaskan North Slope in the old US Naval Petroleum Reserve No. 4. There, in the late 1940s and early 1950s, the uncontrolled movement of tracked vehicles in summer associated with early well drilling activities led to considerable trenching and thermokarst on account of the ice-rich subsoils. In places subsidence along vehicle tracks has formed trenches as much as 1m deep and between 3 and 5 m wide. Large areas of the North Slope are permanently scarred by these tracks. Furthermore, the tracks favour continued thermokarst by collecting water and, if located upon a slope, promote gullying by channelling snowmelt and surface runoff.

In Canada, an unfortunate error was made in 1965 when a summer seismic line programme was undertaken in the Mackenzie Delta. Approximately 300 km of seismic line were bulldozed and long strips of vegetation and soil, approximately 4·2 m wide and 0·25 m thick were removed. Thermokarst subsidence and erosion by running water subsequently transformed many of these lines into prominent trenches and canals over much of their length (Kerfoot, 1974). A more recent example of extensive vehicle track disturbance in Canada occurred during the summer of 1970 on the Sabine Peninsula of Eastern Melville Island. At that time a blow-out occurred at a wildcat well being drilled in the Drake Point area and vehicles were moved, of necessity, across tundra. Where sensitive tundra lowland underlain by soft ice-rich shale of the Christopher Formation was traversed, substantial and dramatic trenching occurred. (Fig. 9.4D).

In general, however, severe erosion associated with vehicle tracks is rare and one must conclude that, for the most part, vehicle tracks present primarily aesthetic rather than terrain problems (French, 1978b, p. 48). Usually, vehicle operators prefer to traverse gently sloping terrain which does not provide the necessary gradient for subsequent deep gullying. Also, the differential settlement of the ground by thermokarst subsidence creates an irregular surface which precludes the development of an integrated drainage system. Furthermore, the introduction of vehicles equipped with special low pressure tyres, the restriction of movement of heavy equipment to the winter months by both Canadian and

FIG. 9.5 CONSTRUCTIONS ON PERMAFROST

(A) (top left) Old and new buildings exist side by side in the city of Yakutsk in central Yakutia, Siberia. The old buildings, placed directly upon permafrost have experienced subsidence. Note also the frost heaving and tilting of the telegraph poles. (Photograph taken July 1973) (B) (top right) Oblique air view of a recently drilled wellsite in the US National Petroleum Reserve (NPR–4), on the Alaskan North Slope. The operation was carried out on a large gravel pad and no damage to the surrounding ice-rich tundra occurred. Husky Oils (NPR) Operations Ltd., South Simpson wellsite. (Photograph taken June 1977) (C) (bottom left) Modern construction techniques, Yakutsk, Siberia. Piles are inserted into the permafrost and construction takes place above the ground surface. (Photograph taken July 1973) (D) (bottom right) At Inuvik, Canada, buildings are built upon wooden piles and services such as water and sewage are effected by a utilidor system which links each building to a central plant. (Photograph taken July 1969)

Alaskan authorities, and the initiation of various terrain sensitivity and biophysical mapping programmes in areas of potential economic activity, particularly in Canada (e.g. Monroe, 1972; Kurfurst, 1973; Barnett *et al.*, 1977), is minimizing this sort of damage.

9.4 ENGINEERING AND CONSTRUCTION PROBLEMS

The thawing of permafrost and the heaving and subsidence caused by frost action can cause serious damage to roads, bridges, and other structures (Fig. 9.5A). In Alaska following the realization of these problems in the 1940s there was a determined effort made by federal and state agencies to improve construction practices and to document permafrost problems (e.g. Ferrians *et al.*, 1969; Péwé, 1966; Péwé and Paige, 1963). In Canada, where large-scale development projects in permafrost regions occurred slightly later, it was possible to benefit from Alaskan experience.

With respect to construction in permafrost, a number of approaches are available, depending on site conditions and fiscal limitations. For example, if the site is underlain by hard consolidated bedrock, as is the case for some regions of the Canadian Shield, ground ice is non-existent and permafrost problems can be largely ignored. In most areas however, this simple approach is not feasible since an overburden of unconsolidated silty or organic sediments is rarely absent. In the majority of cases therefore, construction techniques are employed which aim to maintain the thermal equilibrium of the permafrost.

The most common technique is the use of a pad or some sort of fill which is placed on the surface (Fig. 9.5B). This compensates for the increase in thaw which results from either the warmth of the structure or the destruction of the vegetation that might have occurred during construction, or both. By utilizing a pad of appropriate thickness the thermal regime of the underlying permafrost is unaltered. It is possible, given the thermal conductivity of the materials involved and the mean air and ground temperatures at the site, to calculate the thickness of fill required. Too little fill plus the increased conductivity of the compacted active layer beneath the fill will result in thawing of the permafrost (Fig. 9.6A). On the other hand, too much fill will provide too much insulation and the permafrost surface will aggrade on account of the reduced amplitude of the seasonal temperature fluctuation (Fig. 9.6B). In northern Canada and Alaska, gravel is the most common aggregate used since it is reasonably widely available and is not as susceptible to frost heave as more finely grained sediments.

In instances where the structure concerned is capable of supplying significant quantities of heat to the underlying permafrost, as is the case of a heated building or a warm oil pipeline, additional measures are adopted. Usually the structure is mounted on piles which are inserted into the permafrost (Fig. 9.5C). An air space left between the ground surface and the structure enables the free circulation of cold air which dissipates the heat emanating from the structure. Other techniques used include the insertion of open-ended culverts into the pad, the placing of insulating matting immediately beneath the pad and, if the nature of the structure justifies it, the insertion of refrigeration units around the pad or through the pilings. The use of urethane matting is particularly useful around high Arctic wellsites since there is often an absence of easily accessible gravel aggregate (French, 1978a, p. 14).

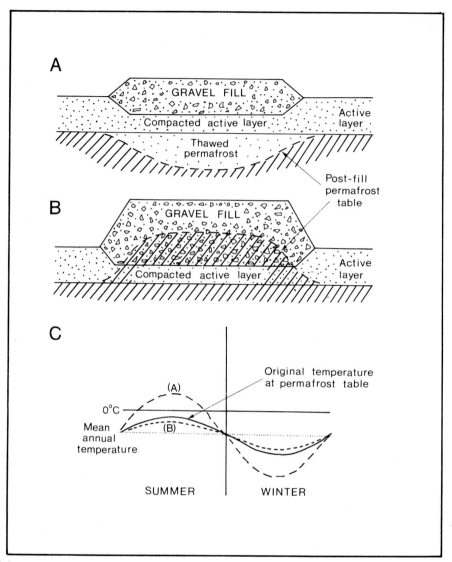

FIG. 9.6 DIAGRAM ILLUSTRATING THE EFFECTS OF A GRAVEL FILL UPON THE THERMAL REGIME AND THICKNESS OF
THE ACTIVE LAYER
(A) Too little fill; (B) too much fill; (C) Effects of cases (A) and (B) upon the thermal regime – too little fill
increases the amplitude of seasonal temperature fluctuation at the permafrost table. (Source: Ferrians et al.,
1969)

The construction of the Alaska Pipeline from Prudhoe Bay on the North Slope to
Valdez on the Pacific coast between 1974 and 1977 utilized many procedures designed to
minimize permafrost problems. First, an access road was constructed along the proposed
route. This consisted of a layer of gravel nearly 2 m thick placed directly upon the tundra.
Working from this pad, heavy equipment enabled the pipeline to be constructed
immediately adjacent to the road. Since the pipe carries crude oil at temperatures which

may exceed 30°C, elaborate measures were taken to ensure that permafrost degradation did not occur. Wherever possible, the pipe was mounted on piles, termed vertical support members or VSMs, inserted into the permafrost. Individual cooling units, using an ammonia solution and capable of airborne monitoring, were mounted on each VSM to conduct heat from the piling to the surrounding cold air. Where the pipeline was buried, the pipe was encased within a specially constructed thermal box surrounded by 25 cm of insulation materials.

In Canada, the construction of the town of Inuvik in the Mackenzie Delta in the early 1960s is another example of the careful manner in which large-scale construction projects have to be undertaken in permafrost regions. A major factor governing the location of the town was the presence of a large body of fluvioglacial gravel a few kilometres to the south (Brown, 1970). Rear-end dumping of these gravels was an essential prerequisite to any heavy construction activity. The result was that the entire townsite was developed upon a gravel pad. Today, the gravel deposit has been exhausted and future growth of the community is dependent upon the exploitation of more distant aggregate sources with their associated higher costs of haulage. The provision of municipal services such as water supply and sewage disposal are particularly difficult in permafrost regions. Pipes to carry these services cannot be laid below ground beneath the depth of seasonal frost as is the case in non-permafrost regions, since the heat from the pipes will promote thawing of the surrounding permafrost and subsequent subsidence and fracture of the pipe. At Inuvik, the provision of these utilities has been achieved through the use of utilidors – continuously insulated aluminium boxes which run above ground on supports and which link each building to a central system (Fig. 9.5D). The cost of such utilidor systems is high, involving a high degree of town planning and constant maintenance, and can only be justified in large settlements.

In the Soviet Union, where modern cities with populations greater than 100 000 have developed in recent years in permafrost areas, utilidor systems constructed mainly of wood or cement are widely employed. In Yakutsk for example, many of the recently completed high-rise buildings are connected to a central large utilidor system for sewage which runs beneath the main street and eventually empties into the large Lena River. As in North America, modern construction takes place on pilings, and airspaces are left between the buildings and the ground surface.

Frost heaving of the seasonally frozen zone is the second major engineering problem to be encountered in permafrost regions (Ferrians et al., 1969, p. 17). Differential heave can cause structural damage to buildings. Equally important is the fact that frost heaving affects the use of piles for the support of structures. While in warmer climates the chief problem of piles is to obtain sufficient bearing strength, in permafrost regions the problem is to keep the pilings in the ground since frost action heaves them upwards. Since heaving becomes progressively greater as the active layer freezes, it follows that the thicker the active layer the greater is the upward heaving force. In the discontinuous permafrost zone, where the active layer may exceed 2 m in thickness, frost heaving of piles may assume critical importance. In parts of Alaska, for example, old bridge structures illustrate very dramatically the effects of differential frost heave (Fig. 9.7). In these regions, it is not uncommon for a thawed zone to exist beneath the river channel. Thus, piles inserted in the stream bed itself experience little or no frost heave. Likewise, piles inserted within

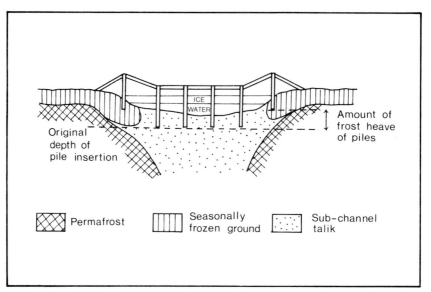

FIG. 9.7 DIAGRAMMATIC ILLUSTRATION OF HOW FROST HEAVING OF PILES INSERTED IN THE LAYER OF SEASONAL
FROST CAN RESULT IN BRIDGE DEFORMATION
(Source: Péwé and Paige, 1963, after an Alaskan example)

permafrost on either side of the river are unaffected since the adfreezing of the permafrost
to the piles provides a resistance to the upward heaving of the seasonally frozen zone. By
contrast, the piles adjacent to the river bank experience repeated heave since they are
located in the zone of seasonal freezing. As a result, uparching of both ends of the bridge
may occur.

In order to prevent heave, piles are usually inserted to a depth of at least 5 m in the
permafrost. In the case of bridges, alternate structures involving minimal pile support are
often considered. In other cases, frost heaving can be minimized by improving the
drainage conditions at the site.

9·5   HYDROLOGIC PROBLEMS

The groundwater hydrology of permafrost regions is unlike that of non-permafrost regions
since permafrost acts as an impermeable layer. Under these conditions the movement of
groundwater is restricted to various thawed zones or taliks. These may be of three types.
First, a supra-permafrost talik may exist immediately above the permafrost table but
below the depth of seasonal frost. In the continuous permafrost zone, supra-permafrost
taliks are rare. In the discontinuous permafrost zone however, the depth of seasonal frost
frequently fails to reach the top of the permafrost since the latter is often relic and
unrelated to present climatic conditions. In these areas, supra-permafrost taliks are
widespread and may be several metres or more thick. Secondly, intra-permafrost taliks
are thawed zones confined within the permafrost and thirdly, sub-permafrost taliks refer
to the thawed zone beneath the permafrost.

Given these hydrologic characteristics, a difficult problem in many permafrost regions is the provision of water to settlements. Since supra-permafrost water is subject to contamination and usually small in amount, and intra-permafrost water is often highly mineralized and difficult to locate, the tapping of sub-permafrost water is vital. In the discontinuous permafrost zone, opportunities exist for groundwater recharge. In parts of Alaska and the Mackenzie Valley extensive alluvial deposits provide an abundant source of groundwater. In Fairbanks, houses rely on numerous small diameter private wells, (Péwé, 1966, p. 28). In parts of Siberia, the occurrence of perennial springs fed by sub-permafrost water assumes special importance to man since these may be the sole source of water available over large areas. In areas of continuous permafrost which may exceed several hundreds of metres in thickness, drilling is either not possible since the hole would freeze or too costly. In these areas, surface water bodies, particularly those which do not freeze to their bottoms in winter, must be utilized and great care taken to prevent contamination. It follows that the supply of water is a severe limitation to any large-scale permanent settlement in the continuous permafrost zone.

A different group of hydrologic problems relate to the formation of icings. These are sheet-like masses of ice which form at the surface in winter where water issues from the ground. Icings are of great practical concern as regards highway and railway construction and in fact, are a distinct hazard to any construction activity. These problems are most common in the discontinuous permafrost zone. Although sub- and intra-permafrost waters may be involved, the most frequently occurring icings are those associated with supra-permafrost water. A common occurrence is where a roadcut or other man-made excavation intersects with the supra-permafrost groundwater table (Fig. 9.8A). Seepage occurs and a sheet of ice forms, often several tens of square metres in extent. In North America, icings were first encountered on a large scale during the building of the Alaskan Highway (Thomson, 1966). Unless precautions are taken, icings can occur on most northern highways which traverse sloping terrain (Brown, 1970, pp. 109–11). Counter measures to reduce icing problems include the avoidance of roadcuts wherever possible, the installation of culverts to divert water from the source of the icing, and the provision of large drainage ditches adjacent to the road. Icings may also block culverts placed beneath road embankments and, by diverting meltwater, initiate washouts in the spring thaw period.

In certain instances, man-made changes to the hydrologic and thermal regime of the supra-permafrost zone can lead to the growth of seasonal icing mounds or frost 'blisters' at locations where they would not occur under natural conditions. For example, Everdingen (1978, p. 275) describes a situation which occurred during the construction of the Dempster Highway in the winter of 1973–4 in the vicinity of Fort McPherson in the Northwest Territories of Canada (Fig. 9.8B). In order to cross a shallow depression which effected subsurface drainage from a nearby lake, an embankment was constructed. The weight of the fill reduced the transmissivity of the water-bearing material of the underlying supra-permafrost talik. In addition, and as a consequence of the fill, permafrost began to aggrade beneath the fill. Thus, subsurface drainage beneath the embankment became restricted. During the winter following construction, therefore, and as a result of the build-up of hydraulic potential in the water-bearing layer, a triangular area upslope of the embankment was uplifted. Repeated ruptures of the frozen soil released flows of water which contributed to a large icing which eventually extended across the highway.

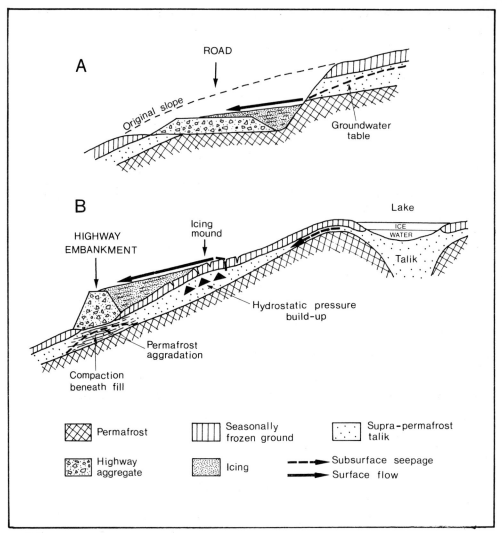

FIG. 9.8 ROAD BUILDING AND PERMAFROST
(A) Schematic diagram illustrating the location of the supra-permafrost talik, and the formation of an icing which might result from a road cutting. (B) Schematic diagram illustrating how a road embankment might lead to the formation of an icing blister. (After an example described by Everdingen, 1978, along the Dempster Highway, Canada)

9·6   CONCLUSION

Increasingly, man's activities are being directed towards the more remote northern regions of the world underlain by permafrost. It can be demonstrated that permafrost, with its associated terrain, ground ice, and hydrologic characteristics, exerts a dominant influence over man's activities, and poses unique geotechnical problems. At the same time, man is able to seriously disrupt the sensitive equilibrium of permafrost environments, probably with more long lasting, costly, and devastating results than in any other environment. A major challenge for the future will be to minimize the deleterious effects of man's activities upon the geomorphic processes of these regions.

REFERENCES

BABB, T. A. and BLISS, L. C., 1974, 'Effects of physical disturbance on Arctic vegetation in the Queen Elizabeth Islands.' *Jnl. Appl. Ecol.,* II, pp. 549–62.

BARNETT, D. M., EDLUND, S. A. and DREDGE, L. A., 1977, 'Terrain characterization and evaluation from Eastern Melville Island.' *Geological Survey of Canada,* Paper 76–23.

BROWN, J. 1966, 'Massive underground ice in northern regions.' *Proc. Army Science Conference,* 14–17 June 1966, Washington, DC (Office, Chief of Research and Development, Department of the Army), I, pp. 89–102.

BROWN, J., RICKARD, W. and VIETOR, D. 1969, 'The effect of disturbance on permafrost terrain.' *US Army CRREL, Special Report 138.*

BROWN, R. J. E., 1970, *Permafrost in Canada; its influence on Northern Development* (University of Toronto Press).

CZUDEK, T. and DEMEK, J., 1970, 'Thermokarst in Siberia and its influence on the development of lowland relief.' *Quaternary Res.,* I, pp. 103–20.

EVERDINGEN, R. O. VAN, 1978, 'Frost mounds at Bear Rock, near Fort Norman, Northwest Territories, 1975–6.' *Can. Jnl. Earth Sciences,* 15, pp. 263–76.

FERRIANS, O., KACHADOORIAN, R. and GREEN, G. W., 1969, 'Permafrost and related engineering problems in Alaska.' *US Geol. Surv. Prof. Pap.,* 678.

FRENCH, H. M., 1975, 'Man-induced thermokarst, Sachs Harbour airstrip, Banks Island, NWT' *Can. Jnl. Earth Sciences,* 12, pp. 132–44.

1976, *The Periglacial Environment* (Longman, London and New York).

1978a, 'Terrain and environmental problems of Canadian Arctic oil and gas exploration.' *Muskox,* 21, pp. 11–17.

1978b, 'Why Arctic oil is harder to get than Alaska's.' *Can. Geog. Jnl.,* 94, pp. 46–51.

HAUGEN, R. K. and BROWN, J., 1970, 'Natural and man-induced disturbances of permafrost terrain,' *Environmental Geomorphology,* ed. D. R. COATES, (State University of New York, Binghamton), pp. 139–49).

HEGINBOTTOM, J. A., 1973, *Effects of Surface Disturbance upon Permafrost,* Report 73–16, Environmental-Social Committee Northern Pipelines, Task Force on Northern Oil Development (Information Canada, Ottawa).

HOPKINS, D. M., KARLSTROM, T. N., et al., 1955, 'Permafrost and groundwater in Alaska.' *US Geol. Surv. Prof. Pap.,* 264–F, pp. 113–46.

KALLIO, A. and REIGER, S., 1969, 'Recession of permafrost in a cultivated soil of interior Alaska.' *Proc. Soil Sci. Soc. of Amer.,* 33. pp. 430–2.

KERFOOT, D. E., 1974, 'Thermokarst features produced by man-made disturbances to the tundra terrain', *Research in Polar and Alpine Geomorphology,* ed. B. D. FAHEY and R. O. THOMPSON, Proc., Third Guelph Symp. on Geomorphology (Geo Abstracts Ltd., Norwich), pp. 60–72.

KURFURST, P. J., 1973, *Norman Wells, 96E/7, map 22; Terrain Disturbance and Susceptibility Maps,* Environmental-Social Program, Task Force on Northern Oil Development (Information Canada, Ottawa).

MACKAY, J. R., 1970, 'Disturbances to the tundra and forest tundra environment of the Western Arctic.' *Can. Geotech. Jnl.,* 7, pp. 420–32.

1972, 'The world of underground ice.' *Ann. Assoc. Amer. Geog.,* 62, pp. 1–22.

McROBERTS, E. C. and MORGENSTERN, N. R., 1974, 'The stability of thawing slopes.' *Can. Geotech. Jnl.,* II. pp. 447–69.

MONROE, R. L., 1972, 'Terrain maps – Mackenzie Valley.' *Geological Survey of Canada, Open*

File Report 125 (maps, scale 1:250 000 of Blackwater Lake, 96B; Norman Wells, 96E; Mahoney Lake, 96F; and Fort Franklin, 96G, map-areas).

MULLER, S. W., 1945, 'Permafrost or perennially frozen ground and related engineering problems.' *US Geol. Surv. Special Report,* Strategic Engineering Study 62 (2nd edn.).

PÉWÉ, T. L., 1954, 'Effect of permafrost upon cultivated fields.' *US Geol. Sur. Bull.,* 989-F, pp. 315–51.

1966, 'Permafrost and its effect on life in the north.' *Arctic Biology,* ed. H. P. HANSEN (Oregon State University Press, Corvallis, 2nd edn.), pp. 27–66.

PÉWÉ T. L. and PAIGE, R. A., 1963, 'Frost heaving of piles with an example from the Fairbanks area, Alaska.' *US Geol. Surv. Bull.,* 1111-I, pp. 333–407.

ROCKIE, W. A., 1942, 'Pitting on Alaskan farms; a new erosion problem.' *Geog. Rev.,* 32, pp. 128–34.

SOLOVIEV, P. A., 1973, *Alas Thermokarst Relief of Central Yakutia,* Guidebook, Second Int. Permafrost Conf., Yakutsk, USSR.

THOMSON, S., 'Icings on the Alaskan Highway.' *Proc. First Int. Permafrost Conf.* (National Academy of Sciences, National Research Council of Canada), 1287, pp. 526–9.

TSYTOVICH, N. A., 1975, *The Mechanics of Frozen Ground* (McGraw-Hill, New York).

TYRTIKOV, A. P., 1964, 'The effect of vegetation on perennially frozen soil.' *National Research Council of Canada Technical Translation,* 1088 (Ottawa), pp. 69–90.

WASHBURN, A. L., 1973, *Periglacial Processes and Environments* (E. Arnold, London).

WILLIAMS, J. R. and VAN EVERDINGEN, R. O., 1973, 'Groundwater investigations in permafrost regions of North America: a review', *Permafrost: North American Contribution, Second Int. Conf. on Permafrost,* Yakutsk, USSR (National Academy of Sciences, Canada) 2115, pp. 435–46.

10

Subsurface Influences

DONALD R. COATES
State University of New York, Binghamton

10.1 INTRODUCTION

Man-induced changes of subsurface materials and processes are highly diverse and cover a broad spectrum of earth science. Depending on the circumstances, these changes may be small or great, slow or fast, and involve disruptions that displace materials, down, up, or laterally. To create the landform modifications the materials may contract, compact, expand, and fracture. The changes that are produced may range from mere nuisances to those that produce disasters with loss of life and property. Total damage inflicted by all human activities that have created subsurface change amounts to billions of dollars throughout the world and probably constitutes the greatest single type of man-related losses.

Only within the last two decades, with minor exceptions, have scientists and engineers started to address problems associated with subsurface influences. Many of the changes occur in the substrate and are not visible, and so, unlike such surface changes as soil erosion, they have received little attention. The changes can be insidious and have such gradual buildup that the damage is already done before it has been recognized. Below-ground changes are also hard to predict, and their prevention and control are even more difficult. The only solution in many cases is the complete abandonment of the endangered area. Awareness and perception of potential problems provide the safest guardians for man's protection, but the number of ways that subsurface materials can be altered mandates a close liaision with scientists and land managers.

Although man lives on the earth's surface, nearly all his activities interact in some manner with the subsurface realm. His buildings may cause settlement, water withdrawn or placed on earth materials may cause their redistribution, underground mining may produce subsidence or collapse, surface excavations can initiate new stresses, and earthquakes can even occur as a result of large water impoundments. There is no easy way to classify the entire range of man's disturbances of the substrate because of dissimilarities in causes and the type of stresses. Indeed results that are produced can be polygenetic in their occurrence.

Whether man deforms the substrate, and what is the character and magnitude of resulting deformation, depends upon many variables. Thus the type, size, and distribution of initiating human activities play an important role in any possible earth modification. In equal manner, the properties of the resisting earth materials help determine their ability to resist alteration. These properties include:

Rock or sediment type, composition, thickness.
Shape, orientation, and arrangement of materials.
Fabric and structure of materials, occurrence of fractures and other planes of weakness, and *in situ* stress conditions.
Environmental setting such as regional and past geologic history of the site, groundwater and other moisture conditions, constraints imposed by boundary conditions of dissimilar physiographic sections.

10.2   LOADING EFFECTS

Man introduces a variety of artificial loads on earth materials which would otherwise be in equilibrium with the geologic setting. These extra burdens have the common denominator of overloading the system by subjecting the substrate to a surcharge of forces. Invariably the pore pressure of the materials is modified, and when the initiating forces exceed the resisting forces, the resulting changes may include compaction of materials, migration of fluids, and rupture of confining rocks. Loading effects are discussed by reference to the consequences of dams and reservoirs, of water injection, of irrigation, and of buildings and other man-made structures.

10.2a   Dams and reservoirs

The first correlation between reservoir filling and earthquakes was made in Greece in 1931 and since that time there have been more than forty substantiated cases of earthquakes being triggered by man-created lakes. The shocks range from microseisms to earthquakes of magnitude 6·4 on the Richter scale. Although to date no dams have failed from this cause, the seismic activity has undoubtedly placed some in jeopardy. Only 0·3 per cent of the world's 11 000 large dams (higher than 10 m) have induced significant seismicity. However, for reservoirs deeper than 90 m, 10 per cent (or 13 dams out of 126) have produced important seismic activity, and of those deeper than 140 m, 21 per cent (4 out of 19) have had significant earthquakes (Mark and Stuart-Alexander, 1977). Most of the sites had several earthquake features in common, namely modest magnitude, shallow focus, maximum shocks related to maximum water levels, and presence of previous tectonic structures although these had been inactive in historic times (see Judd, 1974).

The Marathon Dam, Greece started to impound water in 1929 and earthquakes were first noted in 1931 when the reservoir reached its highest level for the first time. Two damaging earthquakes >5·0 magnitude occurred in 1938. The earthquake history from 1931 to 1966 showed that strongest seismicity was associated with periods of rapid rise in water level.

The filling of Lake Mead behind the Hoover Dam, USA, started in 1935 and the first tremors, in what had been an aseismic area, were felt in 1936. Seismic networks were established in 1937 and during the year about 100 earthquakes were recorded. The maximum event which had a magnitude of 5·0 occurred in May 1939 and significant seismic activity in 1941 and 1942 resulted from rises in water level. However, seismicity has now decreased and there is little relationship of events and changes in water level.

Soon after the major filling of the Koyna Reservoir, India, in 1962, tremors were felt in the areas that had previously been mapped as aseismic. Although five other important earthquakes were felt prior to 10 December 1967, on that date a 6·3 magnitude event occurred that killed about 200 people, injured more than 1500, and left thousands homeless. The city of Bombay, 230 km from the epicenter was shaken severely and the shutdown of the hydroelectric plant paralyzed industry (Gupta, 1976). Other shocks in 1962–73 showed a correlation with water levels in the reservoir. Whenever the water level reached 652 m and remained at that level for some time, then earthquake activity, after some time lag, was increased. Some of the other reservoirs that have produced significant earthquakes include: Monteynard and Grandvale in France; Mangla in Pakistan; Contra in Switzerland; Kariba in Zambia; Kremasta in Greece; Manic in Canada; Hendrick Verwoerd in South Africa; Nourek in USSR; Kurobe and Kamafusa in Japan; Hsinfengkiang in China, and Camarillas in Spain. Although there is some dispute about how the trigger mechanism operates, the consensus by seismologists is that the incidence of earthquakes is related to shear fracturing in rocks. When rocks are already under initial shear stress along an existing fracture plane, an increase in pore pressure caused by fluid migration can be sufficient to overcome the frictional resisting force and to induce shear failure and slippage. Not all reservoirs produce earthquakes because the presence of incompetent strata in the basement rocks of some areas prevents stresses building up, large ambient stress differences are absent in regions not sufficiently deformed, and an active history of tectonism and background seismicity may dampen the slight additional pressure from artificial sources.

The 9 October 1963 disaster at Vaiont, Italy that claimed more than 2000 lives resulted from multiple causes. The immediate reason for the catastrophe was a massive landslide of 300 million m³ of rock that cascaded into the Vaiont reservoir producing a 70 m wave which overtopped the dam and drowned inhabitants in the valley below. Causes contributing to weakening of the rock mass included the presence of dilation joints roughly parallel to valley walls caused by glacial unloading; tectonic rock fracturing; clay minerals in fractures; abnormally high groundwater levels induced by rising reservoir levels; heavy rains in August and September creating a water surcharge in upper slopes; and seismic activity. Reservoir filling started in 1960 and a seismograph installed at the same time recorded 250 tremors during the 1960–3 period, with epicentres 3–4 km from the dam (Gupta, 1976). Water levels in the reservoir were closely correlated with tremor frequency. The three greatest rises in water level were each followed by large bursts of seismic activity. After each peak of activity there was a corresponding decrease in earthquakes.

All of the processes cited probably played a role in weakening the rock mass until the threshold of resistance was exceeded.

10.2b   Water injection

Man-induced earthquakes have also been produced when an abnormal surcharge of water has been pumped into subsurface materials. Starting in 1942, the Rocky Mountain Arsenal of the US Army disposed of chemical warfare products and the waste debris by evaporation from earth reservoirs in the Denver region. However, when it was discovered in 1961 that the wastes were contaminating groundwater and endangering crops, a new method, involving water-tight ponds was tried but this also failed. In March 1962 a new disposal system was inaugurated by pumping the toxic wastes into a 3671 m deep well sunk into Precambrian bedrock consisting of weathered schist and highly fractured hornblende granite and gneiss. In April 1962, Denver, Colorado, felt the first earthquakes, in what had previously been a quiescent area (Evans, 1966). Although the injection programme was halted in September 1965, when proof had been established of the links between the pumping and the seismic activity, the shocks continued for several years and as late as 1969 there were two events of 3·5 magnitude and fourteen events of more than 2·5. In the 1962–7 period, more than 1500 tremors occurred. The magnitudes ranged from 0·7 to 4·3 with most epicentres within 8 km of the well and all within 11 km.

Other water injection schemes are those associated with pumping into aquifers to reverse salt water intrusion in coastal areas, and into petroleum reservoir strata to enhance oil recovery. Only the latter have produced notable subsurface and surface effects. The Baldwin Hills dam and reservoir in California were commissioned in 1951. The site is surrounded by oil wells that constitute part of the Inglewood oil field. In 1954, a pilot water-injection project was so successful in recovering additional oil that an extensive programme began in 1957, and by 1963 there were twenty-two injector wells (Hamilton and Meehan, 1972). The first fault arising from this operation was noticed in May 1957 and by 1963 eight additional ones had been activated, but none were accompanied by recorded tremors. On 14 December 1963, water burst through the foundation and earth dam and in a few hours the 946000 m^3 reservoir emptied onto the communities below. Investigation of the dam failure showed that faults under the reservoir had been activated, had cracked the protective clay liner, had caused 15 cm displacements, and had created small sinkholes in the floor. The dam failure released waters that damaged and destroyed 277 homes, killed 5 persons, and caused $15 million of destruction. The Los Angeles Department of Water and Power brought a $25 million lawsuit against the oil company which was settled out of court for $3875000.

Although costly damages did not result, water injection for secondary recovery in the Rangely field of the Uinta Basin, Colorado, also led to seismic activity. Water injection started in 1957, and in November 1962 a seismological station that was installed 65 km away immediately started recording small earthquakes in the Rangely area. From November 1962 to January 1970 nearly 1000 earthquakes occurred with 320 having a magnitude greater than 1. There was a significant correlation between earthquake frequency and the volume of injected fluid. Earthquakes have also resulted from water injection into the Snipe Lake oil field of Canada (Milne and Berry, 1976) and from hydraulic salt mining operation at Dale, New York.

10.2c   Irrigation

Subsidence caused by application of irrigation waters on loose, dry, low-density soils has come to the attention of researchers only during the past two decades. Large areas in North America, Europe, and Asia contain these materials and especially in the western United States the problems of this type of subsidence have become acute with land lowering of 1–2 m being common, and locally reaching 5 m (Lofgren, 1969). This process of hydrocompaction is especially prevalent in areas where there are either loose, moisture-deficient alluvial deposits ranging from clayey water-laid sands and silts to mudflow materials; or loess and related eolian sediments; or materials which are reasonably fine-grained so that they have a moisture deficiency and the seasonal rainfall rarely penetrates below the root zone. Such deposits under natural conditions have sufficient high dry strength due to clay bonding, cohesion and stacking to support an overburden of a few hundred metres. However, when the dry strength is disrupted by wetting, the materials are forced to adjust to the new pressure system and a different packing arrangement is produced with resultant subsidence.

Damages by subsidence from the surcharge of irrigation waters amounts to many million dollars each year. Much of this damage is to irrigation ditches and canals, well casings, roads, pipelines, and houses. The largest affected area in the United States is in the San Joaquin Valley where more than 500 km^2 is undergoing hydrocompaction subsidence. Other areas are in the Heart Mountain and Riverton areas of Wyoming; near

FIG. 10.1 SUBSIDENCE IN SKID ROW AREA OF SAN FRANCISCO, CALIFORNIA
This area is located on former wetlands that were filled with muds and silts from San Francisco Bay to provide space for urbanization. This house has settled about 2 m as shown by the entrances on both sides of the door, and the windows which were formerly the second story. (Photograph: Don Doehring)

Billings, Montana; on many alluvial fans near Phoenix, Arizona; several areas within the Missouri Basin; and also near Pasco, Washington.

10.2d Buildings and structures

The weight of man-made structures of all varieties may produce settlement of the substrate when improperly engineered. Thus buildings, streets, dams, landfills, canals, etc. may all produce subsidence in underlying materials. (Figs. 10.1, 10.2). The classic case of subsidence consequent upon a building is the Leaning Tower of Pisa, Italy whose tower is about 5 m tilted from the vertical. The tower rests on 4 m of clayey sand which is underlain by 6·4 m of sand that rests on brackish clay (Legget, 1973). In the fifty year period after the completion of the Washington Monument, Washington, D.C., in 1880, the base settled 14·6 cm. In 1913, a concrete grain elevator outside Winnipeg, Canada, with a 0·6 m thick cement slab foundation was completed. The entire structure weighed more than 20 000 tons and was 31 m high, 59 m long, and 23 m wide. After being filled for the first time, 0·3 m of settlement occurred within an hour after movement was first noticed. Within twenty-four hours, the structure was tilted from the vertical by 26°53′ with the west side 7·2 m below its original position, whereas the east side had risen 1·5 m. Legget (1973) also discussed settlement following building in Ottawa, Canada, and Rotterdam in the Netherlands. Zaruba and Mencl (1976, pp. 234–82) provide examples

FIG. 10.2 SETTLEMENT OF HOUSE IN PITTSBURGH, PENNSYLVANIA AREA
(Photograph: Jesse Craft)

of how engineering geology can assist in the structural design of man-made features and so help to prevent and to control settlement of the surface.

One of the largest artificial fills occurs in the San Francisco Bay area of California where one-third of the original bay has now been in-filled. Filling was initiated in 1849. The fill in the Market Street area settled almost 3 m from 1864 to 1964. Typical problems that have resulted include the tilting and settling of buildings (Fig. 10.1) below street level, the cracking of walls, the vertical separation of buildings, and ground sinking around piling foundations (Griggs and Gilchrist, 1977). Ramps have had to be constructed to gain entrance to some buildings and at some bridges more than 6 m of asphalt is needed to counteract localized settlement.

10.3 WITHDRAWAL EFFECTS

Underground extraction of materials is made for a variety of reasons and the resulting subsurface influences, as well as those evident on the surface, are increasingly a world-wide problem. One type of extraction is the mining of natural resources, both solid and fluid, such as coal, groundwater, steam, gas, oil, and salt. The other type of extraction is for engineering purposes such as the construction of tunnels (Attewell and Farmer, 1974). The withdrawal process produces cavities of different sizes and the mining and construction activity may lead to the production of earthquakes, to surface subsidence and collapse, to land uplift, as well as causing other dislocations in materials and in groundwater flow regimes.

10.3a Groundwater mining

Excessive pumping rates that exceed groundwater recharge produce a lowering of the water table and lead to a condition referred to as 'groundwater mining'. The world's largest area of subsidence caused by groundwater withdrawal is the San Joaquin Valley, California where more than 13 500 km^2 have been affected. The average land lowering exceeds 1 m and one 112 km long area has subsided more than 3 m with a maximum of nearly 10 m. The total volume of subsided material in the valley amounts to 186 km^3. The subsidence began in the 1920s when extensive use of groundwater for irrigation was initiated and it increased until the mid-1950s when the annual rate of subsidence was 0·55 m per year on the west side of the valley (Poland and Davis, 1969). The rate of land surface lowering thereafter decreased and was 0·33 m per year in 1963–6. By 1973, the rate had become negligible due to importation of surface water from northern California via the California Aqueduct. Water levels have now recovered by as much as 60 m.

The Santa Clara Valley, California, is another heavily subsided area where groundwater withdrawal for irrigation produced 2·4 m of subsidence between 1934 and 1967 (Poland and Davis, 1969). Groundwater mining started as early as 1916 and increased from $4·9 \times 10^6$m^3 to $18·9 \times 10^6$m^3 per year in the early 1960s. During this time, the artesian pressure head fell more than 75 m and the total volume of subsided sediments amounted to $3·5 \times 10^8$m^3. As is usual in all major subsidence areas, millions of dollars were spent realigning canals and ditches, building up roads, and renovating buildings. Damage to wells has amounted to more than \$4 million and \$9 million in construction costs were required to bolster levees along the south shore of San Francisco Bay to prevent flooding.

The Houston-Galveston region of Texas has also experienced excessive groundwater pumping in the past forty years. By 1943, the subsidence was centred in the suburb of Pasadena and an area 29 km long and 13 km wide had been affected by surface lowering of more than 0·3 m. The rate of lowering increased after 1943 and by 1954 had affected an area double the size and this area had surface subsidence of more than 1·2 m. In 1973, the maximum subsidence was 2·7 m and an area 97 km in diameter had lowered more

FIG. 10.3 ACTIVE FAULT IN PARKING LOT OF ELLINGTON AIR FORCE BASE, HOUSTON, TEXAS
This fracture and similar ones in the area are caused by extraction of fluids, probably mostly oil and gas at this site. (Photograph: Charles Kreitler)

than 0·3 m with surface fracturing (Fig. 10.3). At that time, the pumping rate had reached 1·9 million m³ per day (Spencer, 1977). The extensive damages include coastal and tidal flooding where 2·4 m of freeboard has been lost (Fig. 10.4). Well casings protrude throughout the area and the annual losses in private property exceed $30 million

FIG. 10.4 HOUSE IN WHAT IS NOW THE EXTENSION OF GALVESTON BAY, BAYTOWN, TEXAS
This area has subsided 2·4 m as the result of excessive groundwater production. (Photograph: Charles Kreitler)

FIG. 10.5 ABANDONED WATER WELL IN BAYTOWN, TEXAS
The top of the concrete platform was originally at ground level, showing subsidence due to groundwater withdrawal of 1·5 m. (Photograph: Charles Kreitler)

in a 2450 km² area (Fig. 10.5). A $25 million class action suit brought by citizens against one water company is still pending in the courts.

Numerous other areas in the United States and throughout the world have also been affected by groundwater mining. These include Savannah, Georgia; the Eloy-Picacho area, Arizona; Las Vegas, Nevada; and Denver, Colorado. The rapid growth of Mexico City has led to abnormal groundwater withdrawal (Fig. 10.6). The population expanded from less than half a million before 1895 to 1 million in 1920, to 5 million in 1960, and to 10 million in 1975. Water is obtained from sand and gravel aquifers which are separated by clay and silt deposits and it is pumped from depths of 60 to 500 m. Maximum subsidence now exceeds 7 m. In London, England, a decline in artesian head began about 1820. It had declined 7·5 m by 1843 and by 1936 had lowered by as much as 100 m. Maximum subsidence from this cause is about 0·21 m. High pumping rates in Japan have caused subsidence below sea level in areas where 2 million people live in Tokyo and where 600 000 live in Osaka.

All of the forementioned groundwater subsidence cases occur in sediments that are almost entirely unconsolidated and uncemented. The cause of subsidence is generally attributed to a reduction in the fluid pressure or artesian head. This produces an increased

FIG. 10.6 GUADALUPE SHRINE, MEXICO CITY
The structure has differentially tilted and subsided into the substrate. Less compaction has occurred from the underlying lake beds on the left side of the photograph because bedrock is closer to the surface so that the sediments are thinner. (Photograph: Don Doehring)

load on the skeleton fabric of the granular material. When the bearing pressure cannot be sustained, then intergranular adjustment within the sediment occurs resulting in compaction. A pressure differential exists between the sands and the clays whereby the clay pore wells has amounted to more than $4 million and $9 million in construction costs were required to bolster levees along the south shore of San Francisco Bay to prevent flooding.

Under certain conditions, it is possible for *land uplift* to occur instead of subsidence in heavily pumped groundwater areas. In Arizona, water level decline from irrigation pumping in the 1915 to 1972 period amounted to 48 m in the Lower Santa Cruz basin and 42 m in the Salt River Valley basin (Holzer, in press). Locally, decline of the water table in excess of 100 m was also common in both basins. First-order levelling surveys of the two regions established that during the period 1948 to 1967, there was 6·3 cm of uplift in the Lower Santa Cruz area and 7·5 cm in the Salt River Valley. This uplift, or rebound, occurred in a 8070 km^2 area where $4·35 \times 10^{13}$ kg of groundwater had been removed. The major areas of uplift correspond to areas where crystalline bedrock is close to the surface or crops out through alluvium that was dewatered. The cause of uplift is attributed to elastic expansion of the lithosphere when the groundwater load was reduced or depleted. It is interesting to note that the magnitude of uplift is compatible with the magnitude of depression caused by Lake Mead when allowance is made for differences in the environmental setting. Prior to the groundwater withdrawal, the Arizona region had been one of surface subsidence.

10.3b   Oil and gas production

Withdrawal of fluid hydrocarbons is different from groundwater mining because the reservoir is invariably in rocks rather than sediments and the area of influence is smaller. When the appropriate conditions prevail, such hydrocarbon fluid extraction may lead to faulting, earthquakes, and subsidence. Although there was subsidence from groundwater withdrawal in the Wilmington–Long Beach area of California as long ago as 1928, significant land lowering did not start until major oil production begain in 1938 (Mayuga and Allen, 1966). The first important subsidence of 0·4 m was measured in 1940 and by 1945 had increased to 1·4 m. By 1951, the annual rate of subsidence had reached 0·6 m per year and was causing enormous damage to buildings, pipelines, railroads, and roads. Extensive diking, filling, and other engineering methods were used to counteract the land lowering that ultimately reached about 10 m. Water injection into the strata, after permissive California legislation had been passed, helped to stabilize the area. In addition to subsidence, earthquakes and faulting have also occurred in the Wilmington oil field. Eight separate periods of seismicity have been associated with the oil production and the faults have severely damaged hundreds of producing wells with slippage as much as 22·8 cm during a single seismic event. Total damages from all subsurface influences is much in excess of $100 million (Prokopovitch, 1972).

The earliest recognition of the subsidence-fluid withdrawal relationship occurred in 1925 in the Goose Creek Oil Field, Texas, when levelling showed subsidence had affected an area 6·5 km long and 3·9 km wide with a maximum depression of more than 1m. The subsidence area closely corresponds with the extraction area. In addition, faulting has occurred and earthquakes have been recorded. Some ruptures are 700 m long and many form steeply dipping displacements that are localized along the margin of the subsidence bowl (Yerkes and Castle, 1976). Subsidence in the oil fields of Lake Maracaibo, Venezuela, was first discovered in 1933 and reached more than 3·3 m in some areas by 1954. The clearest example of extraction-induced seismicity by gas outside North America is from the Po Delta, Italy where production of methane gas in 1951 caused a series of earthquakes. Subsidence in an area 40 km long and 20 km wide has created extensive damage from flooding which has necessitated construction of higher levees and drainage from the flooded lands. Subsidence of significant proportions has also occurred from methane gas wells at Niigata, Japan (Poland and Davis, 1969). The general model used by investigators to explain hydrocarbon extraction effects is strata compaction by loss of fluid support. In the rigid units, low angle thrust faults form in the central area with normal faulting along the periphery.

10.3c Mining of solids

The underground extraction of rocks and minerals produce larger cavities than withdrawal of fluids. Thus their effects are likely to be more localized and when ground lowering occurs it is more commonly by collapse of the overburden than by subsidence (Fig. 10.7). Such hazards can lead to disasters where there is total property damage and loss of lives.

FIG. 10.7 SINKHOLES FORMED BY COLLAPSE FROM UNDERGROUND COAL MINING NEAR BEULAH, NORTH DAKOTA (Photograph: John Conners)

Other tragedies occur when the overlying rocks are carbonates. In 1960, a major dewatering programme was initiated in the Far West Rand Mining District near Johannesburg, South Africa so that the gold mines could be extended farther underground (Foose, 1967). This resulted in the formation of some of the world's largest man-induced sinkholes in the carbonates and overburden. Between December 1962 and February 1966, eight sinkholes larger than 50 m in diameter and deeper than 30 m were formed. The largest of the catastrophic sinkholes was 125 m in diameter and 50 m in depth. It occurred without warning. Other sinkholes did extensive damage in the Carletonville area. In 1962, twenty-nine lives were lost in the collapse of a 30 m deep sinkhole and in 1964 five more were killed at another collapse site. During the pumping operations, groundwater levels declined from about 100 m below the surface to more than 550 m in the vicinity of the mines. Springs throughout the area also became dry.

Sinkholes can also be produced during dewatering operations for deep surface mining. To extend their limestone quarries to lower depths in the Hershey Valley, Pennsylvania, the Annville Stone Company pumped an average of 20 m³ per minute during the 30 August to 4 September 1948 period lowering the water level 10 m (Foose, 1953). In May 1949, a new pumping operation discharged groundwater at a rate of 24 m³ per minute and caused a decline in water levels of more than 50 m. This set in motion a chain

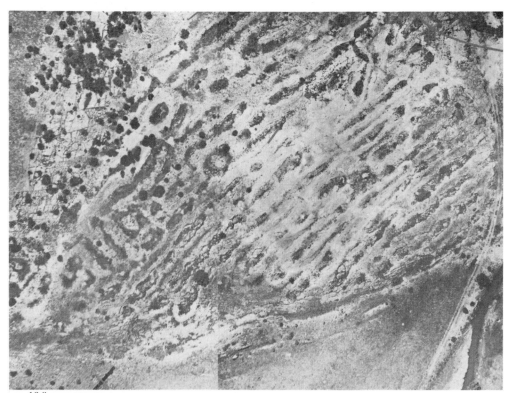

FIG. 10.8 EXTREME SURFACE DISRUPTION OVER MINED-OUT COAL BEDS NEAR MARSHALL, COLORADO
Room-and-pillar arrangement clearly depicted by the collapse pattern. Area viewed is about 470 m long. North is to upper left, parallel to galleries. (Source: Hansen, 1976)

reaction so that water became unavailable for irrigation and crops failed. Springs throughout the area dried up, and during the second month of pumping sinkholes began to form. They ranged in size from a 0·3 to 6 m diameter and with depths of 0·6 to 3 m. The greatest number of the nearly 100 sinkholes formed in the valley and walls of Spring Creek, whose lower course became dry. An interesting sidelight to these conditions was the reaction taken by the Hershey Chocolate Corporation. Because of the dangers being produced by the mining operation, the Corporation started a groundwater recharge programme to raise water levels. They were in turn sued by the Annville Stone Company who sought an injunction to stop the recharge programme. However, both the local courts and the Supreme Court of Pennsylvania ruled that the Hershey Corporation was within its rights to take preventative action against continuing damages that were threatening the community.

In the United States, 28 000 km² of surface land has been undermined in search of minerals and fossil fuels, and 3000 km² exhibit subsidence problems. The greatest hazard is in urban areas where 7 per cent of the subsidence has occurred (Figs. 10.8, 10.9). Coal is by far the largest contributor to the problem. In 1968 extensive damage to streets, bridges, water, gas lines, homes, and churches occurred near Wilkes-Barre, Pennsylvania. Collapses in Coaldale, Pennsylvania damaged or destroyed twenty-three homes in 1963. In Scranton, Pennsylvania, subsidence had become such a severe problem that by 1964

FIG. 10.9 RAILROAD TRACKS DISTORTED BY GROUND SHIFT OVER MINED OUT COAL DEPOSIT, WELD COUNTRY, COLORADO
(Source: Hansen, 1976)

improvements valued at $413 million were necessary in a 0·33 km² area. These Pennsylvania losses have caused that State to pass the Bituminous Mine Subsidence and Land Conservation Act of 1966 (Vandale, 1967). This has helped to prevent new cave-ins by regulating coal mining, mandating sufficient support for all structures, requiring permits and bonding, creating inspection of facilities, and imposing liability damages. In most instances, nearly 50 per cent of coal becomes unmineable because of the necessity for supporting pillars thus preventing more than 100 million tonnes from being mined. Underground coal mining is also a problem in the western United States, such as at Rock Springs, Wyoming. Here the mines range from 3 to 90 m below the ground. Subsidence began to occur in January 1968, has affected several parts of the city, and has damaged homes, streets, and utilities (Fig. 10.10).

FIG. 10.10 SUBSIDENCE FROM UNDERGROUND COAL MINING AT ROCK SPRINGS, WYOMING
Notice washboard character of street, displacements in sidewalk, irregular gutters and fracture offsets. (Source: Geological Survey of Wyoming.)

Underground coal mining can also initiate an entire sequence of events such as occurred in the North Fork of the Gunnison River, Colorado. Mining started in the 1930s to provide coal for a nearby power plant, but coal production ceased in 1953 after methane gas and water were encountered in quantities too costly to control. After the mine was sealed, methane rose to the surface through subsidence fractures in the overlying materials, killing scrub oak and all other woody plants on the surface. In addition, springs ceased to flow in the nearby canyons because the mining had caused diversions in underground flow. Two coal beds below the mine are also threatened as mineable reserves because of the intrusion of methane and water from above, and the overlying coal beds are threatened because of the subsidence fractures.

In addition to subsidence (Fig. 10.11) and collapse, disruption of water flow regimes, and release of harmful gas, there are other influences from subsurface mining. These can

FIG. 10.11 SUBSIDENCE CRACK CAUSED BY UNDERGROUND COPPER MINING AT SAN MANUEL, ARIZONA (Photograph: Allen Hatheway)

be grouped into three categories: namely rockbursts, which are violent failures of explosive character; bumps, which are less violent rock changes; and outbursts, which involve the rapid release of gas that can eject rock with damaging effects (Cook, 1976). Microseismic activity can be associated with all these events. An early example of the relation between the incidence of rockbursts and the rate of mining was demonstrated by the mines of the Brezove Hoty District, Czechoslovakia. Here the annual frequency of rockbursts during the 1913–38 period closely coincided with the amount of rock extracted. Similar relationships between damage and mining activity have been reported in the nickel mines of Sudbury, Canada, the gold mines in Kolar, India, the gold mines in the Witwatersrand District, South Africa, and the zinc mines in the Coeur d'Alene District, United States. Such forces can produce seismic magnitudes as high as 5·0 , and in 1971 there were 1600 tremors produced in the Witwatersrand that ranged 2·0 to 4·2 in magnitude.

Salt mining, both underground and brine pumping, can lead to disasters. Because of poor mining practices and high extraction ratios, collapse has been a frequent occurrence in salt mines near Cheshire, England (Bell, 1975). Mining started in the eighteenth century but few mines lasted longer than forty years before a 'rock pit hole' occurred. The last catastrophic collapse was of the Adelaide Mine in 1928. Brine pumping operations have now induced subsidence as far as 8 km from the site of the producing well and create what are called 'flashes'. These are waterfilled linear hollows developed in the overlying sediments. They may be 10 m in depth and 70 m wide, and they form by collapse of the brine runs. The Cheshire Brine Subsidence Compensation Board was formed because of the many associated hazards and the damage to buildings, farmland, piped services, and road and rail communications. Although some damage claims have been settled, it is often difficult to assess responsibility because effects generally occur at some distance from the producing well which is responsible.

Underground mining for other rocks and minerals has also produced changes in the substrate. Gypsum mining at Garbutt, New York, has created subsidence and sinkholes throughout the mined area, and sandstone mines in the north-east part of Prague have led to subsurface subsidence endangering buildings in the area.

10.4   SURFACE EXCAVATION EFFECTS

Whenever man cuts into the ground, rearranges landforms, or alters surficial water and drainage, changes can be produced in the surface and subsurface materials. These modifications result because the equilibrium that they formerly possessed has been stressed by the imposed man-produced environment. Depending on the character of the dislocation and the type of earth materials, they may expand, contract, or shear.

10.4a   Construction for roads, buildings, resources and services

Piping is a severe problem when excavations are made in some fine-textured and loosely-compacted soils such as loess or lacustrine beds. Aghassy (1973) described piping that evolved prior to 1948 in the southern coastal plain of Israel. These below-surface conduits and passageways ultimately led to a man-created badland topography. The Bedouins used camel-drawn ploughs that incised deep furrows in the loess and led to subsurface infiltration producing vertical pipes that connected with underground horizontal pipes. This produced 2–3 m of collapse in surface materials during a single cycle. Additional badland topography by piping formed as a result of construction of the Ottoman Railroad prior to the First World War. Along the rail route, incision and side-drainage channels created oversteepened slopes. Percolating waters developed steeper gradients with the creation of pipes and the subsequent collapse of overburden into gullies and rills. Many roads on the desert basins in the western United States have also led to piping adjacent to, and under, the roadbeds. Coates (1977) described a piping system that was caused by road construction near Binghamton, New York, which contributed to landslide problems (Figs. 10.12, 10.13). In the USSR, construction of the Khodza-Kola Canal caused piping and resulted in subsidence every kilometre of the 20 km long excavation.

Construction activities in permafrost terrain can also produce undesirable effects when improperly engineered (Fig. 10.14). In Fairbanks, Alaska, the cutting of trees and brush has caused a lowering in the permafrost table of 3·8 m in ten years and consequent subsidence of the ground surface (Cooke and Doornkamp, 1974). It is not unusual for roads to subside as much as 2 m when no allowance is made for permafrost conditions. Several hundred million dollars in extra construction costs were necessitated in building the Trans-Alaska Pipeline because of potential damage that would result from dislocations produced by permafrost processes.

FIG. 10.12 LANDSLIDE IN FINE-GRAINED GLACIAL SEDIMENTS NEAR BINGHAMTON, NEW YORK
The roadcut oversteepened the hillslope and with increased gradients percolation water and drainage from upper road initiated piping. (Source: Coates, 1977)

FIG. 10.13 VIEW OF PIPING IN THE GLACIAL SEDIMENTS OF FIGURE 10.12
This site is in the oversteepened exposure immediately upslope from the lower road. Mattock is 45 cm long and piping conduit extends about 6 cm into hillside. (Source: Coates, 1977)

When excavation is made into bedrock, allowance should be made for possible rock readjustments because of the relief in pressure on the floors and walls of the excavation. Depending upon the amount of *in situ* stress within the rocks, the lithologic composition, and the character of bedding planes and fractures, it is possible for excavations to create microfaulting. Rock creep and swell can amount to many centimetres. Popups and rockbursts also can occur during quarry operations. Sinkholes pose problems in road structure in limestone terrain (Newton, Copeland, and Scarbrough, 1973).

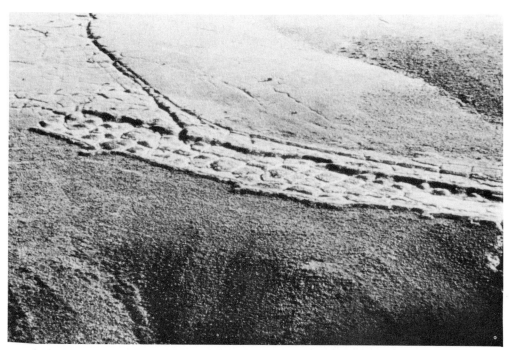

FIG. 10.14 SUBSIDED AND ENTRENCHED MAN-MADE TRAIL NEAR UMIAT, ALASKA
Destruction of the vegetative mat caused disturbance in the permafrost and compaction with desiccation. The
20–30 m long polygons developed in a 10-year period. (Source: Haugen and Brown, 1971)

10.4b Reclamation

Drainage of organic-rich sediments for reclamation purposes invariably produces subsidence by the desiccation–dehydration process. The peaty fens region bordering the Wash on England's east coast has been the site of reclamation drainage for about 400 years (Allen, 1969). At Holme Post after the first pumps for drainage were installed in 1848 the peat sank 1·5 m during a twelve-year period and then sank another 0·9 m in the next eight years. Total subsidence by 1932 was 2·7 m and the original peat thickness has been reduced by desiccation from 6·7 m to 3·4 m. Prior to drainage, where silt sediments cropped out they were 2·1 m lower than the peat exposures. After drainage, the silt areas stood 3 m higher due to difference in subsidence rates and differentials in dewatering. Subsidence has continually been associated with reclamation projects in the Netherlands and other north European areas where marine and delta lands have been reclaimed.

In the United States, the Everglades comprise the largest single area of organic soils in the world covering 7770 km^2. Lowering of the water table by surface drainage began in 1906 and caused volume losses in the peat through biochemical actions under aerobic conditions. All of the soil shrinkage has occurred by oxidation above the water table. The surface subsided more than 1m in thirty years and after forty years there had been a 40 per cent loss in soil volume. The delta confluence of the Sacramento and San Joachin Rivers, California is the second largest peat area in the United States. Here more than

1000 km^2 of peat lands have been reclaimed and cultivated. This area was above sea level before the start of reclamation 100 years ago. By 1953, most of the land surface was more than 3 m below sea level, and the protective levees have now created a saucer-shaped island in the marine environment. New Orleans, Louisiana, is in a wetland area with most land below sea level. An extensive levee system has been engineered to prevent flooding. The city must continually operate water pumps to discharge rainfall runoff and groundwater seepage from the land. The pumping system is the largest in the world composed of twenty-one stations with daily capacity to pump 9.5×10^6m^3 of water (Wagner and Durabb, 1977). Such pumps must keep drained the 222 km^2 of Orleans Parish and 10 km^2 of Jefferson Parish. The US Army Corps of Engineers have recommended a reclamation plan to expand the urbanized area. The project would cost $327 million and it is predicted that during the first fifty years of drainage the peaty soils will subside 3·6 m. The project continues to be in the planning phase.

10.5   OTHER MAN-INDUCED CHANGES

Although not central to the theme of this chapter, there are many additional subsurface-surface side effects that result from man's activities. In coastal areas, excessive groundwater pumping of fresh water has permitted saline intrusion by marine waters into the aquifers. This has caused severe problems in New York, Florida, Texas, and California. Wells that tap geothermal steam sources occasionally produce subsurface influences in California and Iceland. Indeed many landslides result from subsurface influences initiated by man. Font (1977) and Mathewson and Clary (1977) discuss landsliding in Texas caused by highway construction which produced disturbance of the overconsolidated shales that formed the boundary of the excavation. In California, watering of lawns and septic tank effluent have infiltrated subsurface sediments, and this surcharge and the lowered friction of the clayey substrate has created landslides. Even man-made fill structures have failed when sufficient preventative measures for piping were not taken. The Teton Dam, Idaho catastrophe was caused by piping. Waters flowing from the collapsed dam killed eleven, destroyed 120 km^2 of farmland, and caused total damage in excess of $400 million. By June 1977, $168 million had been paid by the federal government in claims. A dam at Lee, Massachusetts failed from piping on 24 March 1968 and killed two people and accounted for $10 million in damage. Even the improper placement and planting of trees too near to buildings can cause their subsidence and cracking of walls because of root growth or as a result of desiccation in some soils.

10.5a   Expansive sediments and rocks

Problems related to expansive sediments (soils) and rocks, were not seriously recognized by soil engineers until the latter part of 1930. Previous to this, damage to buildings and roads was generally attributed to faulty construction and materials (Figs. 10.15, 10.16). The US Bureau of Reclamation first investigated the soil swelling problem in 1938, and since then damages have been related either to settlement or to expansion of the soil. The increasing use of concrete slab-on-ground construction methods after 1940 has further

FIG. 10.15 SCENE IN WACO, TEXAS SHOWING DISRUPTIONS PRODUCED BY EXPANSIVE MATERIALS WHEN DISTURBED BY MAN
The Eagle Ford Shale is over-consolidated and when modified either by removal of overburden or by increased moisture infiltration, it can cause destructive effects. Note undulating character of lawns, fence, and driveway, and the collapsed hard surface road in upper right of photograph. (Photograph: Robert Font)

enhanced the structural damage by expansive soils. Annual damages in the United States linked to expansive soils is now estimated to be more than $2255 million (Chen, 1975), which exceeds loss from such hazards as floods and hurricanes. Principal damage occurs in soils with appreciable montmorillonite clay content. Complete saturation of the material is not needed — moisture increase of only 1 to 2 per cent is sufficient to produce deleterious effects on adjacent buildings or overlying structures. The following solutions can be used to remedy or prevent such losses:

Replace clay soils with non-swelling types.
Provide sufficient deadload pressure on all footings and slabs to withstand cracking.
Flood the area prior to construction.
Decrease the density by compaction.
Change clay properties by chemical treatment and injection.
Isolate the soil so that no moisture change can occur.

FIG. 10.16 RUPTURE AND DISPLACEMENT OF CONCRETE WALL IN WACO, TEXAS
This damage was caused by expansion of the Eagle Ford Shale. (Photograph: Robert Font)

10.5b   Dynamite and nuclear explosions

Detonation of explosives within the earth can cause earthquakes and may produce damage to structures and create other subsurface changes. In the United States, numerous claims are made by property owners who contend that their wells have lost production or their walls have been cracked by explosions from dynamite used in highway and building construction. Nuclear testing can create earthquakes of large magnitude. For example, a 1 megaton explosion may cause tremors of about 7·0 magnitude. At the Yucca test facility in Nevada, movement along one pre-existing fault has occurred for distances of 600 m. In 1968, a nuclear device was exploded underground and activated an earlier fault which opened within seconds after the explosion and caused maximum vertical displacement of 4·5 m.

10.6   CONCLUSION

Man-induced subsurface changes mount each year and the damages increase. The expansion of activities into areas with fragile soils, as in semiarid and permafrost terrain, and the construction of buildings, roads, and reservoirs impose additional stress on the substrate. Knowledge of these changes has lagged because the disciplines of rock and soil mechanics have only developed in the past three decades and earthquake monitoring has only recently reached high levels of sophistication. There is a plethora of ways by which man changes the subsurface environment, but they can be grouped under the main categories of first, placing additional loads on or within the earth; secondly, removal of underground rocks and fluids; and thirdly, surface excavation and drainage. These forces produce a wide variety of responses which include land subsidence or collapse, earthquakes, faulting, expansion or contraction of materials, land uplift, altered groundwater or surface water flow systems, sudden bursts of material, release of gas, and the creation of new topographies. These changes have killed people, destroyed dams and buildings, ruined wells, changed alignments of roads, canals, and ditches, and caused soils to lose productivity.

Multidisciplinary efforts are necessary to prevent or control damage by subsurface changes. Engineers need to be continually aware of the various feedback mechanisms that operate to produce chain reaction effects. This requires perception and investigation by a variety of earth scientists and the enlightened planning by land managers.

REFERENCES

AGHASSY, J., 1973, 'Man-induced badlands topography', *Environmental Geomorphology and Landscape Conservation, Vol. 3 Non-urban regions*, ed. D. R. COATES (Dowden, Hutchinson and Ross, Stroudsburg, Pa.), pp. 124–36.

ALLEN, A. S., 1969, 'Geologic settings of subsidence', *Reviews in Eng. Geol.*, 2, ed. D. J. VARNES and G. KIERSCH, (Geol. Soc. America, New York), pp. 305–42.

ATTEWELL, P. B., and FARMER, I. W., 1974, 'Ground disturbance caused by shield tunnelling in a stiff, overconsolidated clay.' *Eng. Geol.*, 8, pp. 361–81.

BELL, F. G., 1975, 'Salt and subsidence in Cheshire, England.' *Eng. Geol.*, 9, pp. 237–47.

CHEN, F. H., 1975, *Foundations on Expansive Soils* (Elsevier, Amsterdam).

COATES, D. R., 1977, 'Landslide perspectives', *Landslides*, ed. D. R. COATES, Geol. Soc. America, Reviews in Eng. Geol., 3, pp. 3–28.

COOK, N. G. W., 1976, 'Seismicity associated with mining.' *Eng. Geol.*, 10, p. 99–122.

COOKE, R. U. and DOORNKAMP, J. C., 1974, *Geomorphology in Environmental Management*, (OUP, London).

EVANS, D. M., 1966, 'Man-made earthquakes in Denver.' *Geotimes*, 10, pp. 11–18.

FONT, R. G., 1977, 'Engineering geology of the slope instability of two overconsolidated north-central Texas shales', *Landslides*, ed. D. R. COATES, Geol. Soc. America, Reviews in Eng. Geol., 3, pp. 205–12.

FOOSE, R. M., 1953, 'Groundwater behavior in the Hershey Valley, Pennsylvania.' *Bull. Geol. Soc. America*, 64, pp. 623–46.
    1967, 'Sinkhole formation by groundwater withdrawal: Far West Rand, South Africa.' *Science*, 157, pp. 1045–8.

GRIGGS, G. B., and GILCHRIST, J. A., 1977, *The Earth and Land Use Planning* (Duxbury Press, North Scituate, Massachusetts).

GUPTA, H. K., 1976, *Dams and Earthquakes* (Elsevier, Amsterdam).

HAMILTON, D. H., and MEEHAN, R. L., 1972, 'Ground rupture in the Baldwin Hills.' *Science*, 72, pp. 333–44.

HANSEN, W. R., 1976, 'Geomorphic constraints on land development in the Front Range Urban Corridor, Colorado', *Urban Geomorphology*, ed. D. R. COATES, Geol. Soc. America Special Paper, 174, pp. 85–109.

HAUGEN, R. K. and BROWN, J., 1971, 'Natural and man-induced disturbances of permafrost terrane', *Environmental Geomorphology*, ed. D. R. COATES (State Univ. New York, Binghamton), pp. 139–49.

HOLZER, T. L., in press, 'Elastic expansion of the lithosphere caused by ground-water depletion.' *US Geol. Surv.*, preprint.

JUDD, W. R., (ed.), 1974, 'Seismic effects of reservoir impounding.' *Eng. Geol.*, 8.

LEGGET, R. F., 1973, *Cities and Geology* (McGraw-Hill, New York).

LOFGREN, B. E., 1969, 'Land subsidence due to the application of water', *Reviews in Eng. Geol.*, 2, ed. D. J. VARNES and G. KEIRSCH, pp. 271-303.

MARK, R. K. and STUART-ALEXANDER, D. E., 1977, 'Disasters as a necessary part of benefit-cost analysis.' *Science*, 197, pp. 1160–2.

MATHEWSON, C. C. and CLARY, J. H., 1977, 'Engineering geology of multiple landsliding along 1–45 road cut near Centerville, Texas', *Landslides*, ed. D. R. COATES, Geol. Soc. America, Reviews in Eng. Geol., 3, pp. 213–23.

MAYUGA, M. N. and ALLEN, D. R., 1966, 'Long Beach subsidence', *Engineering Geology in Southern California*, ed. R. LUNG and R. PROCTOR (Assoc. Eng. Geol.), pp. 281–5.

MILNE, W. G. and BERRY, M. J., 1976, 'Induced seismicity in Canada.' *Eng. Geol.*, 10, pp. 219–26.

NEWTON, J. G., COPELAND, C. W. and SCARBROUGH, W. L., 1973, 'Sinkhole problem along proposed route of Interstate Highway 459 near Greenwood, Alabama.' *Geol. Surv. Alabama Circ.*, 83, pp. 19–37.

NORMAN, J. W. and WATSON, I., 1975, 'Detection of subsidence conditions by photogeology.' *Eng. Geol.*, 9, pp. 359–81.

POLAND, J. F. and DAVIS, G. H., 1969, 'Land subsidence due to withdrawal of fluids', *Reviews in Eng. Geol.*, 2, ed. D. J. VARNES and G. KEIRSCH, pp. 187–269.

PROKOPOVICH, N. P., 1972, 'Land subsidence and population growth.' *Twenty-fourth Int. Geol. Cong. Proc.*, 13, pp. 44–54.

SPENCER, G. W., 1977, 'The fight to keep Houston from sinking.' *Civil Eng.*, 47, pp. 69–71.

VANDALE, A. E., 1967, 'Subsidence—a real or imaginary problem?' *Mining Eng.*, 19, pp. 86–8.

WAGNER, F. W. and DURABB, E. J., 1977, 'New Orleans: the sinking city.' *Env. Comment*, June 1977, pp. 15–17.

YERKES, R. F. and CASTLE, R. O., 1976, 'Seismicity and faulting attributable to fluid extraction.' *Eng. Geol.*, 10, pp. 151–67.

ZARUBA, Q. and MENCL, V., 1976, *Engineering Geology* (Elsevier, Amsterdam).

Pedosphere

11

Soil Profile Processes

S. TRUDGILL
University of Sheffield

11.1   INTRODUCTION

There is a cynical joke in some planning circles that the best soils for motorway construction are those which occur in disorganized, working class areas! Presumably, soils existing in areas with organized middle class anti-motorway pressure groups are less suitable.

However jaundiced these comments may be, they do perhaps reveal a fundamental truth. Human factors influence the nature of land use; physical factors are not absolute limiting or determining factors and express themselves through economic factors or social value judgement factors. During most rational land-use planning procedures (Curtis, *et al.*, 1976, Ch. 14), the physical potential of a soil is assessed together with other physical factors and an evaluation of the land is made according to the flexibility of the soil resources for a number of possible uses. Social and economic priorities are then assessed and thus any land-use decision is the outcome of the balancing of physical, economic, and social factors.

Soil profile characteristics play a large part in the assessment of physical soil resources. A soil profile is a vertical section through a soil body, and it can be divided into a number of horizontal layers, or horizons, which differ from each other in terms of visible soil characteristics (Fig. 11.1; also see Courtney and Trudgill, 1976). Soil profile processes are of interest since they give rise to the basic nature of the soil. Soil profile processes are therefore of concern to society, and are a focus of study for this chapter, since they act to determine the nature of the soil resources upon which society depends. In addition, the use to which a soil resource is put may often have a noticeable effect upon the soil characteristics themselves. This is less true of the more permanent soil characteristics,

Acknowledgement: I would like to thank F. M. Courtney for comments on this chapter in draft form.

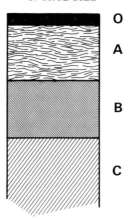

FIG. 11.1 A BASIC SOIL PROFILE WITH HORIZONS
O-organic horizon. A-mixed mineral-organic horizon. B-altered mineral matter. C-unaltered parent material

such as texture and mineralogy, but more true of the *labile* soil characteristics. The term labile is used here to indicate a characteristic which is prone to change under stress and includes such characteristics as soil fertility and soil structure. The study of soil profile processes is thus a two-way one, involving both the influence of soils on man's activities and, in turn, the influence of man on soils. The system which is of interest and which forms the topic of discussion in this chapter is described in outline form in Figure 11.2.

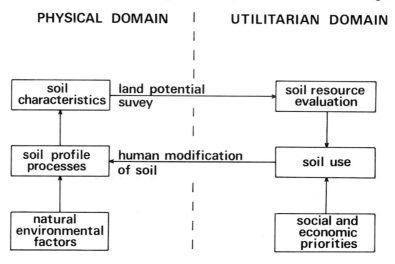

FIG. 11.2 MODEL OF SOIL RESOURCE SYSTEM

Although the diagram (Fig. 11.2) outlines the main connections between the factors involved, there are two other important and related considerations to add. The first is that not only may the external social and economic priorities change, but the social and economic benefit of any selected soil use has to be assessed. This assessment may in turn modify the original economic and social priorities. The second point is that the assessment of the benefit from the selected soil use should involve some consideration of the desirability of the effects of the use on soil characteristics. Thus a continuous reappraisal of the situation often has to be made. This social reappraisal may be related to shifting social priorities and changes in economic conditions. Some land uses preclude alternative uses permanently or in the long term. In other contexts, for example the changes of agricultural land use from one crop type to another or from arable to dairy farming, short-term changes may be seen in response to external pressures. In terms of a physical reappraisal, special studies of the changes of the labile soil characteristics will have to be made; the land potential survey approach is often inappropriate here because this tends to dwell on the more permanent characteristics of soil and site and also tends not to be repeated once made. Thus, for example, a special survey of soil structure or soil fertility may have to be made before an assessment of the benefit or detriment of a soil use to the soil resource may be made. From such assessments, further recommendations for the modification of soil use may be put forward.

It can be appreciated that there is a dynamic, fluid relationship between physical and human factors. The discussion in this chapter has this relationship as its background context. Certain topics have been selected for discussion: namely soil changes and wartime pressures, soil profiles and land reclamation, soil profiles and pollution, modern farming, coniferous plantations and nutrients, together with a consideration of soil resources and society. In these cases it will be demonstrated how the soil profile characteristics reflect the net result of all the human and physical influences acting upon the soil. For example, the study of the disposition of organic matter, chemical elements, and structures in the overall soil profile reflect the combined actions of biological activity, water movement and agricultural practice (Fig. 11.3). The soil profile used by man thus tends to become a cultural product and to embody the fusion of natural processes with man's activities over time. Moreover, it can be viewed from the point of view of being a record of past and

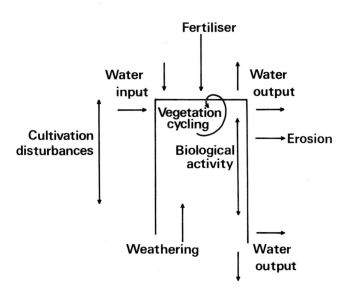

FIG. 11.3 THE MAIN SOIL PROFILE PROCESSES

current natural and cultural pedogenetic processes and from the point of view of assessing the possibilities for potential future uses.

11.2 SOIL PROFILES AND MAN IN THE PAST

11.2a Soils and wartime pressures in Britain

This topic is a classic example of a situation where apparent physical limitations are overcome in the face of economic and social pressures. In Britain during and after the Second World War 'waste', 'difficult', and 'marginal' land was brought under the plough because of the need to increase food production and to place the country on a more self-sufficient footing. Scrub land on heavy clay lands, chalk downland, and steep hillsides were cleared and ploughed.

The effects of ploughing on the soil profile are to redistribute and to homogenise the soil constituents in the plough layer to depths of about 20–25 cm. For example, in the case of chalk soils the uncultivated profile contains a clearly defined 'A' horizon which is dark brown in colour and possesses a high humus content (8–14 per cent organic carbon). However, the ploughed profile is much lighter in colour and more uniform. In this case, the humus is distributed throughout the profile, giving figures of 2–3 per cent organic carbon. The calcium carbonate content of the upper horizons of the unploughed phase may be as low as 15–50 per cent, but in the ploughed phase values of 60–80 per cent occur at or near the surface because chalk has been brought up from the subsoil by ploughing.

11.2b Soil organic profiles on Exmoor

If the study of the modification of soil profiles in the past is extended back, through historical and prehistoric times there is abundant evidence of man's past activities. A study of the evolution of soil profiles over time not only elucidates how present soils obtained some of their characteristics by the activities of man but also, given dateable evidence, provides some information on the rates at which various soil-forming processes operate. This latter type of information can be very important in a management context because it can give some idea of the rates of responses of soil processes to man's activity. From this kind of information it may be possible to predict the long- and short-term effects of alternative soil management policies.

Evidence for the rates at which soil profiles may form can be of two types: buried soil horizons and deductions of soil development from historical evidence. Buried soil horizons may be datable by the study of the pollen present, by Carbon 14 dating or by the observation of the succession of various layers and independent archaeological evidence, such as the construction of Bronze Age barrows over soil surfaces. Secondly, deductions of soil development may be made from a known starting point, for example from areas enclosed from previously unfarmed land (e.g. moorland) or from land reclaimed from the sea at known dates in historical times. In addition, natural events such as landslips and lava flows may provide fresh starting points for soil formation and can be useful in this context if their date of origin is known. Similarly, successive slope erosion events may provide useful information (Curtis *et al.*, 1976, p. 17; Butler, 1959). The evidence leads to the conclusion that from any one given starting point, provided that the conditions are suitable for plant and animal life, the organic fraction of the soil develops relatively rapidly and is quite responsive to changes in external conditions. The response of mineral fractions is somewhat slower (see Trudgill, 1977, Ch. 8 for a general discussion of stability and change in soil and vegetation systems).

The work of Crabtree and Maltby (1975) and Maltby and Crabtree (1976) on Exmoor illustrates several aspects of this type of work. In the study area an iron pan soil was buried by construction activity in 1833, although the original soil was preserved beneath an earthwork. Away from the earthwork soil development continued and also took place on the earthwork itself (Fig. 11.4).

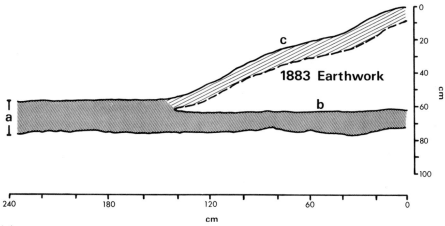

FIG. 11.4 SOIL EMBANKMENT AND BURIED SOIL
(a) uninterrupted development (b) 1883 soil surface (c) soil on embankment. (Source: Maltby and Crabtree, 1976)

Profile studies in 1974 established the extent of soil development over 141 years. In the buried soil the peat horizon is some 12 cm thick, but in the outer area away from the earthwork it is some 40 cm thick. This suggests that about 28 cm of peat has accumulated outside the earthwork in 141 years. From a knowledge of the date of the original field enclosure, calculations suggest that the rate of organic matter accumulation had increased from about 5 to 20 g m^{-2}yr^{-1} before 1833 and to between 50 and 100 g m^{-2}yr^{-1} after this date. The increase is thought to be more probably related to changes in vegetation due to changes in land use, rather than to any other cause such as climatic change.

The nature of the organic matter profile in a soil is an interesting focus of study since it is related to vegetation and land use in terms of the efficiency of biological breakdown processes. Under nutrient-rich situations, biological activity represented by bacterial and fungal decay and ingestion by earthworms, rapidly breaks down organic matter which will be incorporated intimately into the soil profile. Under nutrient-poor or waterlogged conditions biological activity will be curtailed and the organic matter will accumulate as discrete raw humus or peat on the soil surface. Management policies which involve high nutrient supply will thus lead to the absence of a discrete organic surface layer. Thus, it is probable that in this area (Fig. 11.4) a well-managed grazing regime has been relaxed, allowing reversion to a low-nutrient cycling situation. Under high rainfall conditions leaching and waterlogging will be encouraged, leading to greater humus accumulation. In this example, man's influence on soil organic matter accumulation can be seen from evidence deduced from a dateable event which preserved the soil condition existing at a previous point in time.

11.2c Soil mineral profiles on Romney Marsh

Dating evidence of a different kind is provided from a knowledge of the date of origin of a soil parent material. Areas of soil reclaimed from the sea at known times provide examples where the temporal response of different soil profile properties to soil development processes can be readily measured.

An example of the response of the mineral fraction of the soil may be found in the work of Green (1968) on the soils of Romney Marsh in southern Britain. Areas of the marsh have been reclaimed from the sea at various times since at least the ninth century, and the different degrees of soil horizon development on the lands of different ages give good indications of the rates at which various aspects of soil profiles have developed. The primary soil profile processes involved have been decalcification and the destruction of the original sedimentary structures.

Reclaimed land of four main ages has been recognised by Green, and these are Saxon, ninth to eleventh century, fourteenth to sixteenth century, and seventeenth to eighteenth century. It is clear that only soils of similar drainage characteristics should be compared, because in poorly drained soils leaching and decalcification will naturally be slower than in freely draining soils. Data for the topsoil are shown in Table 11.1. The decalcification of the Saxon soils is clearly evident. The calcium carbonate levels of the younger soil also indicates that the process appears to take up to ten centuries to complete.

Table 11.1
DECALCIFICATION OF ROMNEY MARSH SOILS

Type of Land	Calcium Carbonate Content (per cent) of topsoil (0–10cm) of several profiles
SAXON	0·0, 0·0, 0·0, 0·6, 0·0, 0·7, 0·0, 0·0, 0·5 0·0, 0·0, 0·0, 0·0, 0·0, 0·0, 1·8, 0·4, 0·0, 0·3, 0·2, 0·0, 12·6
C9–C11	4·7, 3·6, 4·1, 3·0, 18·8, 14·4, 14·1, 13·4
C14–C16	0·9, 14·3, 3·0, 1·0, 11·9, 17·6, 8·2, 15·6
C17–C18	11·1, 11·4, 15·4, 8·6

Data derived from measurements with a Collins Calcimeter, (the carbonate being released with hydrochloric acid and measured as carbon dioxide given off)

	Mean Percentage	n	Standard deviation
Saxon	0·77	22	2·67
C9–C11	9·51	8	6·28
C14–C16	9·06	8	6·76
C17–C18	11·63	4	2·81

Source: Green (1968)

These data confirm that some mineral portions of the soil tend to respond to changes in external conditions at a rather slower rate than do the organic portions of the soil. This, however, is all based on evidence from the past and illustrates how man's former activities may be related to present-day soil profiles. Having established some idea of the lability of different major soil components, it will now be appropriate to study present-day soil profile processes in relation to man's current or very recent activities. Here, smaller subdivisions of soil components may be dealt with and responses and changes involving, for instance, dissolved and exchangeable nutrients, soil structure, soil pollutants, and soil erosion may be dealt with.

11.3 SOIL PROFILES AND MAN IN THE PRESENT

There is considerable economic and social pressure on modern farmers to produce increasing amounts of food from available soil resources. The general intensification of agriculture has led to the use of larger and more effective farm machinery, the use of pesticides to reduce food losses from pests (especially insects), and the use of herbicides to decrease competition to crops from weeds.

While a social and economic incentive is always very clear, it is possible that soil profile processes may be altered by these activities. Soil structures may be compressed by heavy machinery, and pesticides and herbicides may accumulate in soils. Thus, suggestions have been made that modern farming may harm the soil. Referring to Figure 11.1, it could be suggested that soil use and human modification of the soil may be affecting soil characteristics in such a manner as to decrease the value of the soil resource. The key issue is to evaluate the resilience of the soil with respect to the forces acting on it.

11.3a The soil structure profile

In the case of soil structure, research has shown that some structures are more resilient than others; silt soils seem to be the most prone to compaction and the stability of soil aggregates appears to increase with increasing amounts of soil organic matter. Moreover soil moisture-content plays a very important role in the resistance of structures to compaction. The use of machinery on soils which are wet and above field capacity, when some water surplus to gravity drainage is present, will be far more likely to lead to structural damage than when the same machinery is used on the same soil when drier. Equally, overstocking of pasture with heavy farm animals when the soil is wet can lead to structure deterioration.

Little research has so far been undertaken on the recovery of soil structures to their original form once damaged, but it is clear that a time period of over a year is involved in many cases because patches of bad structure may persist in a field from season to season, even when structures damaged under plough are subsequently left uncultivated. Structure deterioration is illustrated in Figure 11.5. A soil horizon which has been formed by compaction during cultivation comes under the heading of an *agric horizon* in the soil classification used in the USA. It is one example of a specific soil horizon which may be formed in the soil profile through the action of man.

In terms of land-use planning, there are two important management policies involved. First, selection of land for intensive agriculture should avoid those areas where the more permanent soil profile characteristics, such as texture, reveal a propensity to structural deterioration. Secondly, where intensive agriculture is practised, manipulation of the more labile soil characteristics, such as improving drainage and increasing soil organic matter content, will be beneficial. Moreover, cultivation practice which avoids ploughing when the soil is wet is an obvious prerequisite. Examination of the soil structure profile can be

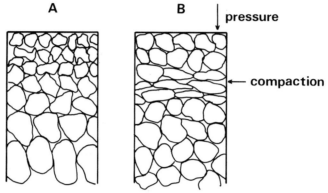

FIG. 11.5 SOIL STRUCTURE PROFILES
(A) a structure profile, (B) a compacted profile

used to indicate whether damage has already occurred. Any subsequent management plan will then naturally have a cost factor built into it. Soils with resilient structures will obviously need no capital investment but soils with unstable structures could be considerably improved by investment in underdrainage and the addition of organic matter. Both these cost money, however, and the likely benefits of increased yield have to be balanced by the costs of the improvement measures. Thus, these kinds of steps will only be taken in an appropriate economic climate where the benefits from higher yield will accrue. Alternative strategies which cost far less are to put the land into use which involves no stress upon the structures, or to carefully manage the timing of cultivation to coincide with occasions when the soil is dry enough, which may not always be possible given the vagaries of weather!

The case of soil structure illustrates well the general themes discussed in the beginning of this chapter. Physical soil profile conditions present a number of possible land-use alternatives. Some soils are more suitable for economically and socially desirable uses than are others. In this case, lack of suitability is to be assessed by a study of structure resilience in the face of intensive physical pressure. The less suitable soils will deteriorate unless costly preventative measures are taken. Rational solution of the problem requires the use of structure stability tests, and the balancing of cost of treatment and of benefit; irrational action involves the use of unsuitable soils for short-term gain, to the long-term detriment of the soil profile and the ultimate devaluation of the soil resource.

11.3b Soil profile pollution: pesticides and herbicides

While soil structure stability may be assessed in part by a survey of soil characteristics, the specification of the overall propensity of soil to pollution is not so easy unless the type and nature of the pollutant is specified. The most important factor is the content of organic matter in the soil. This is both because pollutants may be absorbed into the organic matter and become relatively harmless, and also because organic matter encourages the growth of soil organisms which can act to decompose several soil pollutants.

To proceed beyond the general statement of the significance of organic matter and biological activity, it is often necessary to study the behaviour and sources of specific pollutants in soils in general, rather than the characterisation of soil type in terms of general susceptibility to pollution. Moreover, as far as economic and social evaluation of the problem is concerned, the effects of some pollutants on soil characteristics are usually difficult to see directly; often it is only the side effects which are visible and evaluations may have to be based solely on these. In addition, the differences between the various soil pollutants should be clearly stated.

A most important difference is that between pollutants which can be absorbed and broken down relatively harmlessly and those which persist in and affect soils and organisms. Biodegradeable pesticides and herbicides fall into the former category and persistent pesticides and herbicides, heavy metals, and chemicals which are present in such high concentrations as to alter the soil chemistry fall into the latter. Again, the important issue is the resilience of the soil in the face of external inputs. The resilience in the face of biodegradeable pesticides and herbicides can be assessed with reference to the soil biological population; the resilience in the face of non-biodegradeable pesticides and herbicides and heavy metals with reference to absorption capacity; the resilience to acid pollutants with reference to soil alkalinity (and *vice versa*). It is, thus, very important to specify the nature and type of pollutant involved.

In general, the fate of many modern pesticides and herbicides which have relatively low persistence is divisible into several clearly defined pathways (Fig. 11.6). For instance, in a study of herbicides of the s-Triazine type, Harris *et al.* (1968) stress adsorption, direct volatization to the atmosphere, photodecomposition in sunlight, and decomposition by soil

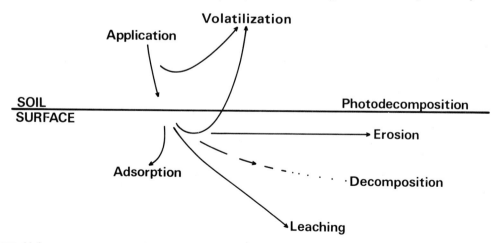

FIG. 11.6 THE FATE OF DEGRADABLE PESTICIDES AND HERBICIDES IN SOIL

organisms. A similar study by Usoroh and Hance (1974) stressed biological control factors (temperature and moisture) in the decomposition of the herbicide linuron in the soil (Table 11.2). Clearly, high temperatures and moist conditions greatly decrease the persistence of the herbicide. The work of Walker (1975) stresses the effects of biological enzyme activity on the persistence and toxicity of pesticides. Leistra and Dekkers (1977) compute the effects of adsorption on the movement of pesticides in flowing water in soils. These topics are reviewed in the works by Kearney *et al.* (1967), Spedding (1975) and in Curtis *et al.* (1976, Ch. 15). The conclusions from all these works are that decomposition of pesticides and herbicides closely follows the factors involved in controlling soil biological activity, and that the disposition of these substances in the soil profile closely follows the pathways of water movement in the soil as modified by retention by adsorption.

Table 11.2
PERSISTENCE OF LINURON (HALF-LIFE) IN A SANDY CLAY LOAM SOIL DEVELOPED ON BOULDER CLAY AND UNDER ARABLE USAGE (COCKLEPARK EXPERIMENTAL FARM, UNIVERSITY OF NEWCASTLE UPON TYNE)

| Water content of soil | Persistence (half-life) | |
Field capacity (per cent)	4°C	22°C
25	41·7	12·5
50	39·4	9·2
100	26·8	8·0

Source: Usoroh and Hance (1974)

Returning to the questions of economic and social priorities, it hardly seems rational to invest money in manipulating soils in order to minimise the detrimental effects of added substances when the inputs of the substances are themselves capable of being manipulated. However, this need may arise if the substances are applied with little thought for side effects. In fact, however, the problem is a decreasing one as biodegradeable substances are being increasingly used. This use is often in response to ecological arguments about persistence of substances in food chains, and the possible damage to organisms at the focus of food chains from persistent substances, as much as it is a response to the needs of agricultural efficiency. Clearly precise dosage of the substances is agriculturally efficient as well as being ecologically more acceptable.

11.3c Soil profile pollution: heavy metals

Pesticides and herbicides are often adsorbed into the lattices of layer silicate clays and in oxides and hydroxides present in the soil, as well as on organic matter surfaces. In these situations the pollutants may be oxidized, hydrolyzed or have their chemical structure broken down by micro-organisms. In solution they may be highly mobile and become involved in hydrological circulation systems and they may also persist in food chains. However, a rather different pattern of behaviour and set of properties is exhibited by heavy metals in soils. These include zinc, lead, cadmium, arsenic, mercury, copper, chromium, selenium, and molybdenum. In the case of these metals there is no decomposition by soil organisms and in many cases the elements are relatively insoluble in the soil and so there is no loss by leaching. Often, the only soil losses of these metals is by the physical erosion of the soil itself. While this lack of mobility is beneficial in that these metals do not readily find their way into water systems in solution, their persistence in the soil profile means that once contaminated, the soil is likely to remain that way.

There are both natural and man-made sources of heavy-metal pollution in soils but the distribution of the problem tends to be of a more local and acute nature than the more widespread pesticide and herbicide pattern. Geological sources include concentrations of individual elements in particular rocks. For instance, the rock serpentine has a high concentration of chromium present. Other sources derive from ores and mining. Thus, soils developed on mining spoil, and on alluvium derived from mining spoil, are particularly prone to heavy-metal pollution. Most heavy metals are more soluble under acid conditions and it is therefore possible to identify acid soil profiles (such as podzols) as presenting a greater potential hazard than alkaline soils. Clearly, a useful management policy for soils polluted with heavy metals is the addition of lime to increase soil alkalinity.

Tackling heavy-metal pollution at source may be possible in some situations. Cadmium may be present in traces in phosphate fertilizers; copper may be present in fungicides; copper and zinc can be found in some prepared feed blocks and mercury may be present in seed dressings. All these substances are applied in agricultural practice and careful monitoring of chemical quality and examination and control of manufacturing processes and inputs can help to minimise the levels of metals present.

In addition, lead can originate from road traffic fumes as well as from mining, and also chemical elements, such as sulphur, can be present in the air from industrial pollution. Sulphur can be a particular problem as in the form of sulphuric acid, it can acidify the soil, unless the soil has a high base status which can act to offset the acidity. Obviously, a heavy-metal pollution problem coupled with a sulphuric acid input can lead to a particularly acute situation.

In terms of management and plant nutrition, some chemical elements are needed as plant nutrients. Copper, zinc, and sulphur are in this category and need to be present in sufficient minimal concentration for satisfactory plant growth. Other elements, such as nickel, lead, cadmium, mercury and chromium are not essential and can be termed pollutants at whatever concentration they occur. Soil profile evaluation for resistance to metal pollution should identify organic rich, alkaline soils as the most useful and mining spoil soil profiles as the ones to avoid. If the source cannot be tackled and the soil has to be pressed into agricultural use, liming is often a useful management policy. The case of heavy metals may, in fact, however, represent one where locally the contamination of the soil profile is so acute as to represent a severe limitation on land-use possibilities. In this case the soil may not be worth treating.

11.3d Soil profile pollution: fertilizers

Policies of increasing agricultural productivity often involve the addition of chemical fertilizers to the soil profile. Much of the added fertilizer finds its way to its intended destination, namely plant root uptake; however, proportions are also retained in the soil and are lost by leaching (Fig. 11.7). The proportions involved in each route vary with the fertilizer involved and with the nature of the soil, but nitrogen is particularly soluble and can be readily leached. In this case, sandy soils, and other soils where soil water movement is rapid, have to be treated with particular care, making sure that minimum doses of nitrogen are applied, otherwise any excess will be quickly lost. Phosphorous, on the other hand, is strongly absorbed in the soil, so much so that it may become unavailable to plants even though present in large quantities. In acid soils the added phosphate may react with hydrous oxides of iron, aluminium and manganese and become fixed and unavailable. In alkaline soils, it may form an insoluble calcium phosphate compound. The retention of

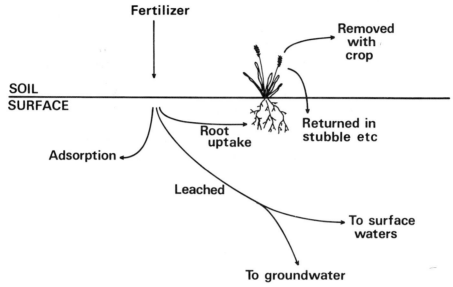

FIG. 11.7 THE FATE OF APPLIED FERTILIZERS IN THE SOIL

phosphate may range from 20 to 60 per cent of that added as fertilizer. This has become such a prominent feature of some soils fertilized by man, especially the USA, that distinct horizons in the soil profile can be recognized as a result. In the United States Department of Agriculture classification of soils, the term *epipedon* is given to the dark, organic rich, surface horizon. An *anthropic epipedon* is recognized, with the word anthropic indicating an origin from the action of man. This surface horizon is a dark, base rich layer, with a high amount of phosphate present which has accumulated over long periods of agricultural use. Such horizons may have over 1000 ppm phosphorous present in them, compared with 200–300 ppm at a lower depth (of about 40 cm). Manipulation of the soil pH to between 6 and 7 may help to maximise phosphate availability as it appears to be least available at extreme pH values.

In these cases, man's actions do not necessarily limit the usefulness of the soil. In the case of phosphate accumulation, no serious effect on soil productivity is seen and indeed, it may be beneficial. In the case of nitrate leaching, the side effects are usually more serious than the effects on the soils. Concentration in inland drainage waters may encourage aquatic plant growth to an unwelcome degree, such that aquatic animal life may suffer. This is more a problem of the agricultural ecosystem than of the soil profile. It illustrates the far reaching effects that soil profile processes may have, and careful management of fertilizer application to the minimum effective doseage is clearly necessary both from the points of view of agricultural economics and of ecological acceptability.

11.3e Soil profiles and conifer plantations

In upland Britain, plantations of coniferous trees tend to be advocated for the soils which are less suited to agricultural use. This is not to say that timber production would not be high on the better agricultural land, it simply reflects the fact that the better land gives a larger and quicker economic return under agriculture. The poorer land often gives better returns under forestry than under agriculture even though sheep farming may also be profitable. The poorer soils tend to be acid and/or organic, infertile, and stony. They are often on steep and rocky slopes or in wet areas.

Controversy has existed concerning the effects of the plantation of conifers on the long-term productivity of the soil. On the one hand, is the argument that tree roots are thought to be able to pull nutrients up from low down in the soil profile where they have been washed by leaching. The nutrients then enter the trees and are recycled to the soil surface during litter fall, thus enriching the soil surface with nutrients. On the other hand, it is pointed out that while the deep-root uptake and recycling of nutrients is true for trees in general, conifer litter is usually acid. This encourages the acidification of the topsoil and the mobilization and leaching of soil nutrients because many of them are more soluble under acid conditions. It is also often pointed out that, compared with many broadleaved trees, many conifers (such as the spruces), are shallow rooting and therefore they do not necessarily have access to nutrients present lower in the soil profile. The evidence for these two viewpoints is conflicting and it is probable that the factors involved vary from site to site. This appears to make it possible to gain evidence both of soil acidification under conifers and of a complete lack of effect, depending upon the conditions prevailing at the site studied.

One important point concerns the nature of conifer litter. Conifer needles contain high amounts of *chelating* substances. These are organic compounds capable of incorporating nutrient cations, such as calcium, in their structure. When chelates are washed from the surface leaf litter horizons to lower layers in the soil they take up and carry with them such nutrient elements, thereby impoverishing the topsoil.

Supporting evidence comes from the work of Handley (1954) who measured the acidity of leaf-water extracts from conifers and broadleaved trees (Fig. 11.8). While there is an overlap between pH 4·1 and 4·8, there is also a clear trend for many conifer species to have more acid leaves than many broadleaved species.

It is clear that it is necessary to specify which species is under discussion as they differ in their relevant characteristics. In addition, soil conditions should also be specified. Wet soils lead to shallow rooting: for example, Douglas Fir usually has a strong, deep tap root, but on wet soils it only tends to develop a horizontal root network at shallow depths. But the most important factor is, again, the resilience of the soil in the face of the input. Soils with limited reserves of bases and nutrients are liable to be affected by an input of acid litter. Those with high reserves will be able to offset, or buffer, the effects of acid litter (though this point is not necessarily relevant as often these are the most fertile soils which, as discussed initially, are not usually planted with conifers). The soils where the least effects will be seen are those which are already acid and have low nutrient levels. Thus Ovington and Madgwick (1957) found that on acid and infertile upland soils in North Wales and in podzolic soils there was very little effect. The order of the pH differences between conifers and hardwoods appeared to be of between 0·2 and 0·3 pH units around pH 5 in the topsoil. Conversely, soils of moderate base status (pH 6–7) have tended to become more acidified, by up to 0·5 pH units. Such an acidification may mean that in the long term, the productivity of moderate base status soils may be decreased.

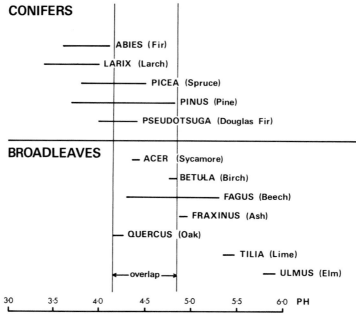

FIG. 11.8 ACIDITY OF LEAVES OF CONIFERS AND BROAD-LEAVED TREES
(Source: Handley, 1954)

The case of the possible acidification of the soil profile by conifers again hinges on the use of a fundamental knowledge of soil profile processes. This is used in order to evaluate which of the alternative uses will cause least long-term detriment to the soil resources involved. In this case, soil resources may be unaffected if the already acid soils are used for conifer plantation; however degradation of soil resources may occur if sensitive, moderate base status, soils are used. Thus, in terms of land-use planning and resource survey, site characteristics of soil acidity, reserves of bases, and susceptibility to leaching will be important soil profile characteristics to evaluate before a rational land-use plan can be drawn up.

11.3f Soil profiles and society

In this chapter, frequent reference has been made to the social importance of the knowledge of soil composition. It is desirable to re-emphasise some of the important points by brief mention of a case study of soil erosion and its social aspects which will link with the topic of the next chapter (Chapter 12).

It has been stressed that man's activities may affect the quality and flexibility of soil resources and that man should use his knowledge of soil profile processes in order to manipulate the soil resource to the maximum benefit of society and also for the maximum conservation of the resource. However, one point which should be emphasized more fully is that the environmental scientist who works on this kind of problem should be one who is fully socially aware of the implications and ramifications of his work. This is simply because soil profile processes are often so closely inter-related with social processes.

Few studies illustrate these points better than that of Blaut *et al.* (1959) on the cultural determinants of soil erosion and conservation in the Blue Mountains of Jamaica. The background to the study was that intensive cultivation, largely of coffee, had led to widespread soil erosion. Truncation of the soil profile was common, with the loss of the fertile topsoil. An initial plan was drawn up for soil conservation measures to be put into practice. The plan was, however, a physical or engineering solution to what was seen as a physical problem of soil profile erosion. The limited success of the plan was due to inattention to social and cultural factors. It was not until factors such as land tenure, social contact patterns, perception of erosion problems, market behaviour, social structure, local pride, and cash flow in the community were examined and constructively incorporated into the plan that soil conservation measures, such as terracing, began to have any degree of adoption and success.

Clearly, the lesson from this is a general one. Application of physically or biologically sensible soil management plans may not always succeed without cognisance of their social and economic context.

11.4 CONCLUSION

In Figure 11.1 the relationships between soil profile processes, soil characteristics, soil use, and social and economic factors are stressed. Some of these relationships have been discussed in greater detail by means of case studies.

It has become evident that many soil profile characteristics may be modified by man and that studies of past modification not only reveal this but also indicate something of the rates of changes occurring. Many of the changes have possible detrimental effects and the concern is that such effects should not limit the value of soil resources for the future. The conclusion from the case studies examined is that spatial surveys of soil profiles and the evaluation of soil resources should be carried out in this context. They should be concerned with assessment of the nature of the soil profile in terms of characteristics relevant to soil profile processes involved in offsetting any possible detrimental effects, for example, high organic content as an absorber of pesticides and herbicides or alkalinity as a decreasing agent of metal toxicity. Moreover, especial attention should be focused upon the resilience and ability for recovery of the soil profile in the face of the inputs induced by man, as discussed by Trudgill (1977, Ch. 8). Equipped with this kind of understanding, soil resources can be manipulated for the long-term benefit of man by the maintenance of soils in a state thought to be the most productive for a selected use, without any detriment to the soil. Clearly the allocation of a soil resource to its most suitable use is a challenging field involving both physical and social evaluation. It is one worthy of attention since the rational and beneficial allocation of limited land resources is one of the difficult and pressing tasks that faces present and future society.

REFERENCES

AVERY, B. W., 1964, 'The soils and land use of the district around Aylesbury and Hemel Hempstead.' *Mem. Soil Surv. Gt. Brit.* (Harpenden).

BLAUT, J. M., BLAUT, R. P., HARMAN, N. and MOERMAN, M., 1959, 'A study of the cultural determinants of soil erosion and conservation in the Blue Mountains of Jamaica.' *Social and Economic Studies*, 8, 403–20.

BUTLER, B. E., 1959, 'Periodic phenomena in landscapes as a basis for soil studies.' *Soil Publ. CSIRO, Austral*, 14.

COURTNEY, F. M. and TRUDGILL, S. T., 1976, *The Soil. An Introduction to Soil Study in Britain* (Edward Arnold, London).

CRABTREE, K, and MALTBY, E. M., 1975, 'Soil and land use change on Exmoor. Significance of a buried profile on Exmoor.' *Proc. Somerset Archaeol. Nat. Hist. Soc.*, 119, 38–43.

CURTIS, L. F., COURTNEY, F. M. and TRUDGILL, S. T., 1976, *Soils in the British Isles* (Longman, Harlow).

GREEN, R. D., 1968, 'Soils of Romney Marsh.' *Bull. Soil Surv. Gt. Brit.*, 4 (Harpenden).

HANDLEY, W. R. C., 1954, 'Mull and Mor in relation to forest soils.' *For. Comm. Bull.*, 23 (HMSO, London).

HARRIS, C. I., KAUFMAN, P. D., SHEETS, T. J., NASH, R. G. and KEARNEY, P. C., 1968, 'Behaviour and fate of s-Triazines in soils', *Advances in Pest Control Research*, 8, ed. R. L. METCALF, 1–55.

KEARNEY, P. C., KAUFMAN, D. D. and ALEXANDER, M., 1967, 'Biochemistry of herbicide decomposition in soils', *Soil Biochemistry*, ed. A. D. McLAREN, and G. H. PETERSON (Edward Arnold, London), 318–42.

LEISTRA, M. and DEKKERS, W. A., 1977, 'Computed effects of adsorption kinetics on pesticide movement in soils.' *Jnl. Soil Sci.*, 28, 340–50.

MALTBY, E. M. and CRABTREE, K., 1976, 'Soil organic matter and peat accumulation on Exmoor: a contemporary and palaeoenvironmental evaluation.' *Inst. Brit. Geog., Trans., New Series*, 1, 259–78.

OVINGTON, J. D. and MADGWICK, H. A. I., 1957, 'Afforestation and soil reaction.' *Jnl. Soil Sci.*, 8, 141–9.

SPEDDING, C. R. W., 1975, *The Biology of Agricultural Systems* (Academic Press, London).

TRUDGILL, S. T., 1977, *Soil and Vegetation Systems* (OUP, London).

USOROH, N. J. and HANCE, R. J., 1974, 'The effect of temperature and water content on the rate and decomposition of the herbicide linuron in soil.' *Weed Research*, 14, 19–21.

WALKER, C. H., 1975, 'The persistence and toxicity of pesticides in relation to enzymic activity.' *Outlook on Agriculture*, 8, 201–5.

12

Soil Erosion and Conservation

RORKE B. BRYAN
University of Toronto

12.1 INTRODUCTION

It has been said that if each soil conservationist stopped the movement of one grain of soil for each word he has written on the topic, the problem of soil erosion would disappear. It is true that despite close attention from numerous scientists around the world for the past sixty years, soil erosion is now perhaps more widespread than ever before. At the same time it must be realized that much of the technical knowledge necessary to solve erosion problems has been available for years, but the actual solution requires social, political, or economic incentives which are more difficult to stimulate. Without such initiatives the technical solutions discussed below will have little impact. In soil conservation the way to perdition is liberally paved with the excellent, but ignored, reports of technical consultants.

Soil erosion is a natural geomorphological process essential to landform evolution and the continued release of mineral nutrients by weathering to maintain soil productivity. Our concern is not with natural soil erosion, but with soil erosion accelerated by human mismanagement to a rate sufficient to cause disruption and decline of agricultural productivity. Furthermore, attention will be directed towards the ways in which man may positively modify the physical system to control soil erosion rather than the extent to which human activity can accelerate the process. A review of soil conservation measures provides a useful example of man's positive manipulation of physical processes although, paradoxically, the need to undertake that manipulation frequently stems from his own mismanagement.

The threshold for acceptable annual soil loss depends to some extent on the soil type and depth, cropping patterns and the availability of alternative land, but a figure of 1·8 tonnes per hectare (t ha^{-1}) is commonly used in the United States (Smith and Stamey, 1964.) Erosion may involve entrainment and transport by either wind or water, which are separate and distinct processes although some important controlling factors, such as certain soil properties, may affect both. At the same time wind and water erosion often interact in the same area in a complex cycle of widespread degradation. It is such areas of composite erosion, particularly semi-arid marginal lands in developing countries, which have been the focus of most recent international concern (e.g. Rauschkolb, 1971; Rapp *et al.*, 1972; FAO/UNEP, 1974; Rapp, 1974; Rapp, *et al.*, 1977). In many cases a comprehensive, integrated approach to erosion problems is essential and the separation in the discussion below is merely for convenience.

12.2 WATER EROSION

In the process of water erosion the energy necessary to entrain and transport soil is the kinetic energy of either rainsplash or overland flow (see Chapter 7). Both can entrain material separately, but frequently they interact in a complex manner so that the precise mechanism of detachment is not clear. In the case of rainsplash, raindrop velocity does not vary much and energy variations therefore depend chiefly on drop size, which in turn depends on rainfall type. Drop sizes of rainfall are normally distributed but the median drop size for low-intensity frontal rainfall of lengthy duration is substantially lower than that for high-intensity short-lived convective rainstorms. Accordingly, convective rainfall is a much more potent agent of rainsplash erosion. Whether soil is, in fact, detached depends partly on the presence or absence of protective vegetation, which may absorb most of the impact energy, and partly on the soil character. Rainsplash will not be competent to disrupt and move coherent soil unless it is weakened by saturation, but it can splash discrete particles and aggregates into the air. On a level surface, splashed particles will fall uniformly around the point of impact, but when the surface is sloping, the majority of particles will land downslope in an effect which increases with slope angle (e.g. Ekern, 1950; De Ploey and Savat, 1968). The critical controls on rainsplash erosion are therefore rainfall type and intensity, proportion of ground free of vegetation, soil character, and slope angle.

Material is entrained by overland flow primarily due to shear stress exerted on the surface, although in sheetwash a lift force caused by variations in pressure around individual particles may also be significant. In either case the force involved is closely related to flow velocity, which in turn is governed by the volume of flow, the roughness of the surface and the slope angle. This relationship is usually described by the Manning equation:

$$V = K \frac{R^{2/3} S^{1/2}}{n}$$

where: V=flow velocity; R=hydraulic radius; n=Manning's roughness factor; S=slope angle; and K=a constant depending on units used. For a thin sheetwash the hydraulic radius is almost equal to the depth of flow. As the depth of flow increases with the length of flow, so does the velocity. Initially the flow has no capacity to entrain, but as it accelerates it eventually reaches an entrainment threshold velocity whose magnitude depends on the surface resistance. Where the surface is bare of vegetation, resistance to entrainment depends on soil properties and, for any given soil, the distance required to achieve the threshold velocity will vary with the flow volume, the surface roughness and the slope angle.

The fundamental control on soil erosion by overland flow is rainfall in excess of the soil's infiltration capacity. The infiltration capacity may be limited by the rate at which water can pass through the soil surface or by the total void space available to hold water. In both cases the size, shape, and packing of soil textural separates and aggregates are important controls. Where the surface is of low permeability or the overall water-holding capacity is low, as in shallow desert soils, surface runoff can be generated even by short rainstorms of moderate intensity. Where permeability and water-holding capacity are high, surface runoff is rare, occurring only after prolonged or extremely intense rainfall. In all circumstances, intense rainfall, even if short-lived, is most likely to generate runoff, particularly if soil aggregates are unstable and collapse under raindrop impact to clog pores and create an impermeable surface seal as illustrated in Figure 12.1 (McIntyre, 1958; Bryan, 1973).

In its initial stages overland flow is discontinuous and much disrupted by storage in surface micro-depressions, but if rainfall continues these will eventually be overtopped and a continuous sheet established. Depending on the flow velocity reached and on soil properties, entrainment may occur in this sheetwash. Non-coherent soil resists entrainment primarily by the size and mass of discrete particles, following the relationship established

FIG. 12.1 SURFACE CRUSTING BY WETTING AND RAINDROP IMPACT CAN REDUCE INFILTRATION AND IMPEDE SEEDLING EMERGENCE

by Hjulstrom (1935). Particles may be either textural separates or aggregates and in the latter case the situation becomes more complicated because these may retain their size and shape on wetting, or disintegrate, depending on the strength with which the components are cemented. If the soil body is coherent or if it has become sealed by raindrop impact it resists entrainment primarily by its shear strength. In this case the flow velocity necessary to initiate entrainment is much higher (experimental flume data would suggest by a factor of four or five).

Entrainment may or may not occur in sheetwash, though it is more likely to do so if the sheetwash is disturbed by raindrop impact, which retards flow (Yoon and Wenzel, 1971; Savat, 1977) but splashes soil into suspension. As the sheet moves downslope it starts to concentrate and split into a series of concentrated streams in which flow velocity, and therefore erosive potential, are enhanced. This leads rapidly to incision of narrow deep rills which deepen by coalescence and capture into gullies, distinguished from rills only by the impossibility of obliteration with normal tillage techniques.

12.3   ASSESSING EROSION HAZARD

While much attention in soil conservation is usually directed to remedial action in severely eroded areas, it is clearly preferable to minimize the problem where possible by avoiding areas of highest erosion hazard. Identification of the variables controlling the water erosion process permits a comparatively precise prediction of such hazard.

Rainfall is the dominant and most fundamental component and many studies have been carried out to refine measures of rainfall hazard or erosivity. The difficulty is to find a single measure which encompasses all the influences of rainfall. The most outstanding work is that of Wischmeier and his colleagues (e.g. Wischmeier et al., 1958; Wischmeier, 1959) who used over 10000 plot-years of data from experimental stations operated by the US Soil Conservation Service to isolate an EI_{30} factor as the measure most closely correlated with soil loss. This measure is the product of the kinetic energy of the rainfall (E) and the maximum 30-minute intensity (I_{30}) for an individual storm. Summation of indvidual storm values provides an annual index and these have been used to produce maps of mean annual erosivity values for the United States and Canada which clearly identify Florida and the Gulf Coast as the zone of maximum hazard. The annual index has been refined to provide an assessment of seasonal hazard which permits adaptation of agricultural practice to reduce hazard. Although successful in North America, the EI_{30} factor is not uniformly applicable in other countries. In Africa, Hudson (1971) found the measure (KE>1), based on the total kinetic energy of rainfall above a certain threshold necessary to initiate erosion, to be more appropriate. Clearly, similar refinements could be developed for other areas. In all cases, however, the importance of rainfall intensity in developing an erosivity index is a stumbling block, because comparatively few climatic stations record rainfall intensities, and in any case the storms of highest intensity tend to be so localized that they are easily missed by climatic stations, particularly in sparsely instrumented developing countries.

Precisely the same approach can be employed to arrive at an assessment of the hazard posed by soil properties. The pioneer work on erodibility was that by Middleton (1930) and since then numerous attempts have been made to isolate a single property which can be used as an index of soil erodibility. None has proven entirely satisfactory in comparative testing (Bryan, 1968), but recently certain measures of soil aggregation and aggregate stability have shown sufficient promise to justify further testing (Bryan, 1976). Maps of soil erodibility are of particular significance for they provide an early warning of potential trends in soil degradation before severe erosion starts and while remedial measures can be relatively easily applied.

Mapping of rainfall erosivity and soil erodibility are both potentially valuable in regional land use planning. On the same scale topographic hazard can be assessed by ordinary relief maps. For planning land use at a local or farm level a more refined tool, the Universal Soil Loss Equation has been developed by Wischmeier, viz:

$$A = RKLSCP$$

where: A = soil loss in tons/acre; R = rainfall erosivity based on the EI_{30} factor; K = a soil erodibility factor based on soil loss from a 1/200th hectare plot on a 9 per cent slope; L = a length factor relating the locality to the test plot length of 22·6 m; S = a slope factor relating the locality to the test plot slope of 9 per cent; C = a crop management factor comparing actual crop type to the test conditions of cultivated fallow; and P = a conservation practices factor which compares the agricultural practice to the worst case of ploughing directly up and down the steepest slope (Fig. 12.2). The Universal Soil Loss Factor has been very widely applied, often to areas and situations for which it was never designed (Wischmeier, 1977) and has been criticized because it is clearly less than universal in applicability. Nevertheless, when properly applied it is a useful management tool at the field scale, not least because it provides a means of clearly quantifying the contribution which adoption of a conservation measure can make to reducing soil loss below the tolerance threshold. It also provides a ready means of assessing benefits to be compared with the costs of adopting any measure.

12.4   CONSERVATION PRACTICES

Most of the conservation practices available are designed to protect the complete field area from the combined effect of rainsplash, sheetwash and rillwash, and differ from the specialized measures used to prevent expansion of gully systems. Field area techniques are directed towards four basic objectives:

1 Protection of the surface from raindrop impact
2 Maintenance or increase of infiltration capacity
3 Reduction of overland flow velocity
4 Reduction of soil erodibility.

Many measures also serve the purpose of water conservation, which is at least as important in many of the areas prone to erosion. They may involve either crop management or mechanical protection or commonly some combination of both depending on the intensity of soil erosion.

FIG. 12.2 PLOUGHING PARALLEL TO THE STEEPEST SLOPE ON EVEN A VERY GENTLE GRADIENT CAN LEAD TO RUNOFF CONCENTRATION AND INITIATION OF GULLY EROSION

12.4a Crop management

Crop management techniques may be viewed as an essentially low-level response to soil erosion, involving comparatively little disruption, expert consultation, or cost. Nevertheless, considerable benefits can be derived from crop management as indicated by values for the *C* factor in the Universal Soil Loss Equation which vary from 0·05 to 1, providing a potential for twenty-fold reduction in soil loss. Crop management is also flexible and comparatively easily adapted to annual or seasonal variations in erosion hazard.

Crop management influences soil loss directly by determining the surface area exposed and the period of exposure. It also indirectly affects infiltration, runoff generation and erodibility through its influence on soil physical properties. Surface exposure is particularly significant where rainsplash hazard is high, as in many tropical lands, and careful selection of crop type may reduce soil loss significantly. Crop selection may be limited by market conditions but, even if the crop cannot be changed easily, alteration of the cropping cycle to ensure that a well-defined period of peak rainfall hazard does not coincide with a period of particular vulnerability may be adopted. Hudson (1957) has described an almost 50 per cent reduction in soil loss under tobacco in Rhodesia resulting from early, rather than late, planting. Another approach to the same problem is the practice of intercropping with staggered crops so that a complete field is not simultaneously exposed, crops approaching maturity providing protection for those at seedling stage.

The other crop management practice available to keep the surface covered is stubble mulching, in which the crop stubble or residue is left in place after harvest to protect the soil at the beginning of the following growing season before the crops have reached protective height. The technique was developed originally to protect North American prairie farms against wind erosion, and is still primarily used for that purpose. It can lead to problems with seeding and the emergence of seedlings, and to a proliferation of weeds and soil pests. A related system which is effective without those negative effects is trash farming in which chopped crop residue is spread and ploughed into the soil to produce an improved tilth in the surface soil.

The maintenance of good tilth is the key to high infiltration capacity and low erodibility. Although it is essentially the summation of all soil physical properties, it is most closely related to aggregation. Aggregation is a complex and dynamic property influenced by many factors such as the type and abundance of clay minerals, the cations adsorbed onto the soil colloidal complex, and the abundance and nature of the organic matter content. Little can be done to adjust unfavourable clay mineralogy, but other controls respond well to cropping practice. Essentially, crop management methods which maintain soil at high fertility will also contribute to good aggregation. It is perfectly possible (though increasingly expensive) to maintain soil fertility at a high level by addition of chemical fertilizers, but soils maintained in this way tend to suffer a progressive reduction in organic matter content, which in most soils is essential to the maintenance of stable aggregation. Few attempts have been made to monitor the decline of organic matter, but it is known that on some of the immensely fertile black soils of Alberta, Canada, organic matter has declined by 50 per cent since continuous cropping started at the beginning of the century, and Eckholm (1975) has documented similar alarming declines in montane areas in the Indian sub-continent.

The best means of combating a decline in organic matter is to introduce a suitable crop rotation, preferably one involving a grass crop and a nitrogen-fixing legume such as clover or lucerne, which will combine abundant restoration of organic matter with the beneficial rooting effects of grass. In many areas, particularly in developing countries, intense land scarcity will not permit the abandonment of continuous cropping, and it may ultimately prove necessary to introduce some form of commercial organic fertilizer, perhaps moved from areas of organic surplus such as Canada. Where rotation is possible, the beneficial effects in terms of erosion reduction are unquestionable (Moldenhauer, *et al.,* 1967).

12.4b   Mechanical protection practices

Mechanical protection techniques are designed to reduce soil loss by reduction of overland flow and by preventing overland flow achieving threshold entrainment velocities. A bewildering variety of practices has been developed around the world to achieve this basic objective with varying effectiveness, but not all have necessarily been based on a firm understanding of the scientific principles involved.

One of the oldest and most universal measures is contour tillage or the operation of all farm implements across the slope. In the case of ploughing, this retards concentration of water into channels and therefore the onset of rillwash. Each furrow acts as a dam holding up runoff and increasing infiltration. Contour ploughing is an appropriate measure for controlling erosion caused by moderate intensity rainstorms on gentle slopes. The effectiveness has been demonstrated by many studies such as those of Van Doren et al. (1950), who found that it decreased erosion by 37 to 60 per cent in Illinois. It can cause problems when abnormal storms occur, because the concentrated release of water produced by the overtopping of a contour ridge may enhance erosion. This occasional disadvantage is more than outweighed in most areas by the general benefits which have been recognized at least since Jefferson adopted contour ploughing in 1813. Despite this, an astonishing number of farmers, at least in North America, still plough parallel to the steepest slope. Many practices similar to contour farming which involve only minor adaptations of normal agricultural methods have developed in different countries. In much of East Africa, for example, the practice of tied-ridging is used, in which ridges are thrown up across the furrows, dividing them into small basins. This technique, called basin-listing in the United States, is an effective water conservation method which can improve yields, but, as with contour ploughing, can cause enhanced erosion in severe storms. In Africa the hand-raised contour bund is a slightly higher bank providing more protection than the simple contour ridge, but in India the same term is used for a much larger and more elaborate structure.

The mechanical protection methods described above are sufficient where erosion hazard is moderate but if slopes are steep, rainfall intensities high, or soils particularly vulnerable, then the more elaborate, expensive, and initially more disruptive method of terracing must be adopted. Again this is a very ancient technique, although terraces have been used for different purposes. Many of the early terraces, such as those of the olive groves and vineyards of the Mediterranean basin, were level-floored bench terraces intended only to provide level ground for cultivation, although some were elaborately constructed with stone risers and have survived intact for thousands of years. This construction was extremely labour intensive and therefore such terraces, although still used, are now seldom constructed. Also very ancient and widespread are the elaborate irrigation terraces found throughout much of SE Asia. In this case the terrace has a water-retaining lip, so that water is retained to infiltrate in place on the terrace, as in rice paddies, or is led slowly downslope through the terrace system.

Both bench terraces and irrigation terraces may incidentally help to reduce soil erosion, although this is not their prime purpose. For erosion control the standard method is the channel terrace designed to intercept overland flow and lead it away from the field at non-scouring velocities. The design of such a system is much more elaborate and technically demanding, because the hydraulic characteristics and spacing of channels must be such that they can handle all the runoff supplied from the terrace and remove it rapidly and without scouring. Likewise the terraces must be spaced as far apart as possible, but sufficiently close to prevent overland flow down the terrace reaching entrainment velocity before entering the channel. Fundamental to the complete system is a major stormwater diversion drain which, by intercepting flow originating outside the farm area, will ensure that the terrace system will only have to handle locally generated runoff. This diversion drain is critical, for if underdesigned it will initiate failure of the complete system. The drain leads water laterally, at a slight angle across the contours, to a grassed waterway which removes water rapidly to an established watercourse. This waterway also intercepts runoff from the terrace system and because it will handle concentrated flow at high velocities, it must be lined with a well-established resistant grass sward or, in some cases, with concrete. If this fails it will rapidly degenerate into a gully system which may destroy the complete farm. Once the diversion drain and grassed waterway are established, the terrace system can be developed. The spacing of terraces depends on the slope angle, the infiltration capacity and soil erodibility, and is usually determined by a generalized expression, such as that used by the US Soil Conservation Service who calculate the required vertical interval between terraces as

$$VI = aS + b$$

where: VI = vertical interval in feet; a = factor varying from 0·3 in southern States to 0·6 in the north; S = slope angle (%); b = factor of 1 or 2 depending on soil type.

Detailed design of the terrace varies, but essentially it consists of a shallow channel and low soil ridge which will vary in width from 3 to 15 m. In many systems the complete terrace is cultivated, but in some cases the channel is kept in grass and in others permanent vegetation is grown on the ridge. The channel floor may be parallel to the contour so that runoff collects and eventually infiltrates (absorption terrace), or it may lead at a constant or variable slope (graded terrace) to the grassed waterway. In the latter case the length and slope of the channel and its detailed hydraulic character must be planned so that water does not reach entrainment threshold velocity before it reaches the grassed waterway.

Space does not permit a detailed discussion of the design procedure for channel terraces which is described in a number of texts (e.g. Hudson, 1971). The basic principles involved are accurate forecasting of the volume of overland flow, which is essentially a hydrological problem, and detailed design of drains, waterways, and channels to avoid scouring, which is a problem of hydraulic engineering. Where adequate data are available for forecasting and adequate funds exist, an excellent terrace system can be developed which will reduce soil loss dramatically. Hays and Palmer (1935), for example, quote an 85 per cent reduction in soil loss due to terracing in Wisconsin. For much of the world, data are not adequate, rainfall intensity measurements being particularly deficient. Rainfall of the highest intensity is often very localized and is frequently not recorded by sparse climatic stations. Rapp *et al.* (1972) have documented the effects which localized catastrophic rainfall can have on upland agriculture. Elsewhere the costs involved in terrace construction cannot be supported by the agricultural system, unless labour is very cheap. Finally, even where design and construction are satisfactory, the system must be maintained. The world is littered with the remnants of erosion control systems which have failed due to inadequate maintenance.

12.4c Control of gully erosion

It is rather questionable whether erosion by combined sheetwash and rainsplash or by gullying is more significant. Sheetwash and rainsplash probably remove a greater volume of fertile soil, but gullying, while actually removing less surface soil, can dissect land into minute uneconomic fragments, can totally disrupt the operation of machinery and, by lowering the water table, can reduce yields or pasture growth even where the soil is intact. (Fig. 12.3). Gullying can occur anywhere where water is concentrated and vegetation

FIG. 12.3 GULLIES, WHILE DIRECTLY REMOVING ONLY A SMALL AMOUNT OF SURFACE SOIL, MAY DISSECT AND RENDER USELESS VERY LARGE AREAS OF FARMLAND

cover is incomplete. It frequently originates along tracks and roadways and may be caused by overgrazing or by cattle trampling around water holes. One classic case quoted by Bennett (1939) was started by the drip from the gutter of a barn. Once initiated, gully systems can expand extremely rapidly, and their control therefore often requires drastic measures. The essential stages in gully control are:

1 Temporary or permanent reduction of the volume of water flowing through the gully
2 Reduction of the velocity of flow
3 Stabilization and revegetation of gully banks.

Temporary reduction of the volume of flow is required to allow other work to proceed, whilst permanent reduction may be needed where the catchment area of the gully has grown so large that a stable channel system of adequate capacity cannot be established. Temporary reduction may be produced by diversion ditches, and may be unnecessary where gully flow is ephemeral. Permanent reduction will require modification of the catchment area and possibly construction of small reservoirs.

Measures to reduce flow velocity are usually designed not to bring velocity below entrainment threshholds for bare soil, but to protect re-established vegetation. They frequently involve a variety of small check structures, such as brushwood fences, rip-rap gabions, small brick weirs, old car bodies or other locally available materials, but sometimes quite elaborate small dams may be necessary. Once these structures are in place, stabilization and revegetation may proceed. Stabilization is necessary where banks are extremely steep and slumping may occur, and is required particularly where vertical banks are being undercut by lateral erosion. This may involve grading and bulldozing, and sometimes the use of explosives. The subsequent stage of revegetation may be impeded by the exposure of infertile subsoil, and fertilization is sometimes necessary. Choice of plants is determined by rapidity of growth and root system development, by suitability to local climates and by degree of protection provided.

Whilst the basic processes of water erosion have been fairly clearly understood for some time, there has been a tendency to view and treat them in isolation. The potential effects of this have been well demonstrated by Nir and Klein (1974) in south-west Israel, where adoption of terracing reduced sheet erosion, but the increased infiltration stimulated piping and the enhanced headward development of gullies. This emphasizes the desirability of integrating geomorphological knowledge with agricultural engineering techniques in the solution of water erosion problems.

12.5   WIND EROSION

It is almost impossible to estimate accurately the relative importance of wind and water erosion, but wind erosion can at times occur on an extremely dramatic scale so that public awareness of, and administrative reaction to, soil erosion has been largely stimulated by wind erosion. This was certainly the case in Washington in 1935 when Franklin Roosevelt achieved passage of his Federal Soil Conservation Act against a sky backdrop darkened by clouds of topsoil blown from the Great Plains. The 'dust bowl' years have left an indelible print on the landscape and psyche of the North American plains and certainly were a key factor in widening United States governmental authority in the realm of environmental control, although similar sweeping governmental powers had been granted in India by the Punjab Land Preservation Act of 1900. In the Ukraine the 'black winters' of the late nineteenth century also caused much damage (Yakubov, 1969), while in this decade the plight of the Sahel is only too familiar.

The potential for wind erosion exists anywhere that dry unconsolidated soil is exposed to air movement. Under exceptional circumstances wet sand can blow, but such circumstances are too rare for this to be of agricultural importance. In practice wind speed seldom rises sufficiently high to entrain material larger than 0·84 mm in diameter (Chepil and Woodruff, 1963) or smaller than about 0·008 mm, so wind-erosion hazard depends not only on wind speed, exposure, and drying regime, but also on the availability of suitable soils. As with water, the threshold entrainment velocity necessary to initiate particle movement is higher than that required to sustain it (fluid threshold and impact threshold, respectively (Bagnold, 1941)). Entrainment results primarily from the surface drag of the wind (Bagnold, 1941; Chepil and Woodruff, 1963) although turbulent air currents, electrical charges on particles, and the lift force are also significant. Once entrained, material may be transported well above the surface in suspension, or as a series of bouncing grains in saltation, whilst some material too large to move off the surface may be moved by the impact of falling particles as surface creep. In general the saltation component is by far the most important, accounting for between 50 and 75 per cent of the material moved. Because of the significance of saltation and particle impact and with other factors remaining constant, soil movement tends to increase downwind in an effect referred to as 'soil avalanching'.

Apart from the damage which may be caused by deflation, the abrasive effect of particles in transport can be severe. Chepil and Woodruff (1963) recorded a 78 per cent reduction in wheat forage yield and an 86 per cent reduction in grain yield following ten minutes exposure to a 45 km h⁻¹ wind. This emphasizes the serious damage which may be caused by short periods of high winds. In April 1968, many farmers around Edmonton, Alberta, lost up to 15 cm of topsoil during a week of strong winds.

12.6   PREVENTION OF WIND EROSION

The essential objectives of measures designed to reduce wind erosion are:

1 Maintenance of surface cover
2 Reduction of wind velocity
3 Entrapment of moving soil
4 Increased erosion resistance of the soil.

As with water erosion, remedial or protective measures may involve comparatively inexpensive modifications to standard cropping practice, or expensive and disruptive development of shelter belts.

12.6a   Crop management and tillage

Because wind cannot entrain soil protected by vegetation, any measure which increases the area or duration of cover will be beneficial and these include the practices of stubble mulching, trash farming, and intertilling, described above. Because the entraining force is lateral, not vertical as with rainsplash, intertilling is not very effective unless the crop covered strips are wide, when the practice is known as strip cropping. This was adopted very widely in western North America after the dustbowl. It can be very effective in reducing entrainment or in entrapping soil and can contribute to moisture conservation by trapping snow drifts. The effectiveness depends on the orientation, which should be as close as possible to perpendicular to the direction of maximum wind velocity.

Because of frictional drag, a very thin layer of stagnant air lies immediately adjacent to the ground. Any measures which increase surface roughness increase the thickness of this layer and reduce entrainment. Both strip cropping and stubble mulching are effective in reducing wind velocities in this critical boundary layer, while tillage may produce the same effect. In some cases deep ploughing may be carried out as an emergency measure to provide some protection in a particularly hazardous period. As an example of the effect, Chepil and Woodruff (1963) reported a threefold reduction in soil loss as a result of surface roughening. The tillage implement should be one that leaves the surface vegetation undisturbed, such as the duck-foot cultivator, or one that produces a rough, cloddy surface, such as a lister plough or chisel. Any treatment which disturbs the surface without appreciable roughening, as with discing or harrowing, should be avoided. Apart from emergency measures, tillage should be minimized because of the disruption of soil aggregates and its effect in increasing erodibility. One of the major contributory causes of the dust bowl was the practice of dust mulching carried out in the erroneous belief that a surface layer of pulverized soil would significantly reduce moisture loss by interrupting capillary rise. Current practice is to increase surface aggregation as much as possible by ploughing up well-structured, clay-enriched subsurface soil and by avoiding winter fallowing in which aggregates are broken down by frost action.

12.6b Windbreaks and shelterbelts

Where wind hazard is particularly high, the protection provided by strip cropping and tillage practices should be supplemented by vegetation shelterbelts which absorb wind energy and deflect the windstream, providing reduced velocity to leeward. The degree of protection is affected by the size, shape, orientation, and structure of the belt. Ideally it should be perpendicular to the wind direction, wedge-shaped and have some porosity, but not enough to induce jetting and leeward erosion. The extent of the protection provided depends on the height of the barrier, being typically six to nine times the height. This means that the spacing should be rather close, with fields seldom wider than about 100 m. This is a major disadvantage with shelterbelts for the ideal profile and structure requires a number of rows of shrubs and trees of different height. Bennett (1939) describes an ideal shelterbelt consisting of nine rows which would use over 6000 trees in an 800 m long belt (Fig. 12.4). Not only does such a belt use a considerable amount of land but it can also cause serious moisture depletion in adjacent soils. Furthermore, the cost of trees and shrubs can be prohibitive. In Canada one of the major measures of the Prairie Farms Rehabilitation Act (1935) was to set up government nurseries which provide a wide variety of shrubs free to farmers. Despite this, full-scale shelterbelts are rare in the Canadian prairies, and most of them protect ranch buildings rather then fields.

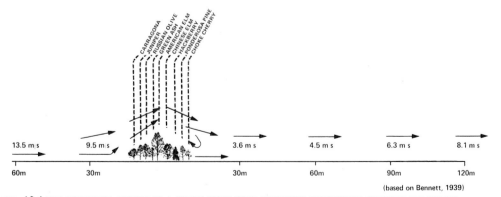

(based on Bennett, 1939)

FIG. 12.4 THE POTENTIAL EFFECT OF A FULLY DEVELOPED NINE-ROW WINDBREAK ON WIND VELOCITIES
Such a windbreak would use over 6000 plants for a 800 m long break, which is not only expensive, but may also lead to moisture depletion in adjacent fields

The measures discussed are essential to prolong successful farming in semi-arid marginal lands where wind erosion is prevalent. In western North America most were widely adopted in the late 1930s with great success. The dreaded scourge of drought and high winds does not occur every year, however, and the memory of the dustbowl has dimmed. In many places strip cropping has been abandoned and shelterbelts have fallen into decay, often destroyed by wind-drift of herbicides used along roadsides. The hazard is still real, however, and a return of the conditions of the early 1930s could again cause extensive damage. Throughout these lands, wind erosion can be controlled but often the cost of conservation techniques, added to those for irrigation, for control of water erosion, for drainage, for fertilization, and for transport, make profitable farming impossible. For some areas, as in North Africa and the Soviet Union, it must be recognized that the only real solution is to abandon cultivation and restore the grasslands. In south-eastern Alberta the population dropped by 50 per cent and the number of communities by 60 per cent during the 1930s, as devastated farms were turned into community pastures. Thoughout this region abandoned clapboard farm houses are a constant reminder of the physical limits of agricultural development.

12.7   CONCLUSION

The discussion of erosional processes has drawn on recent research which has led to refinement of conservation techniques, but it must be stressed that much of the technical knowledge required for erosion control has been available for fifty years. While there is scope for increased understanding, the major challenge is now the development of conservation measures and strategies which conform to the economic, social, and political realities of the countries where they are needed. Further development of remedial conservation practices based on high investment, elaborate machinery, and prolonged expert maintenance is no longer particularly useful. Major progress in soil conservation now requires approaches at different levels. At one level there is a need for the integration of expertise from numerous disciplines to develop conservation practices based on low-level technology and methods of extension which will lead to widespread adoption. At a different level there is a clear need for development of global and national monitoring systems which will provide early warning of soil degradation, and permit initiation of preventive rather than remedial measures.

REFERENCES

BAGNOLD, R. A., 1941, *The Physics of Blown Sand and Desert Dunes* (Methuen, London).
BENNETT, H. H., 1939, *Soil Conservation* (McGraw-Hill, New York).
BRYAN, R. B., 1968, 'The development, use and efficiency of indices of soil erodibility.' *Geoderma,* 1, pp. 5–26.
    1973, 'Surface crusts formed under simulated rainfall on Canadian soils.' *Laboratorio per la Chimica del Terreno, Pisa, Consiglio Nazionale delle Richerche, Conferenze,* 2.
    1976, 'Further considerations on soil erodibility indices and sheetwash,' *Catena,* 3, pp. 99–111.

CHEPIL, W. S. and WOODRUFF, N. P., 1963, 'The physics of wind erosion and its control.' *Advances in Agronomy,* 15, pp. 211–302.

DE PLOEY, J. and SAVAT, J., 1968, 'Contribution a l'étude de l'érosion par le splash.' *Zeitschrift für Geomorphologie,* 12, pp. 174–93.

ECKHOLM, E. P., 1975, 'The deterioration of mountain environments.' *Science,* 189, pp. 764–70.

EKERN, P. C., 1950, 'Raindrop impact as a force initiating soil erosion.' *Soil Sci. Soc. Amer. Proc.,* 15, pp. 7–10.

FAO/UNEP, 1974, *A World Assessment of Soil Degradation, Report of Expert Consultation, Rome, 10–14 June,* (UNEP/FAO).

HAYS, O. E., and PALMER, V. J., 1935, 'Soil and water conservation investigations.' *Soil Conservation Service Progress Report* (Upper Mississippi Valley Soil Conservation Experimental Station).

HJULSTROM, F., 1935, 'Studies of the morphological activity of rivers as illustrated by the River Fyris., *Bull. Geol. Inst., University of Uppsala,* 25, pp. 221–527.

HUDSON, N. W., 1957, 'Soil erosion and tobacco growing.' *Rhodesia Agric. Jnl.,* 54, pp. 547–55.
1971, *Soil Conservation* (Batsford, London).

McINTYRE, D. S., 1958, 'Permeability measurements of soil crusts formed by raindrop impact.' *Soil Science,* 85, pp. 185–9.

MIDDLETON, H. E., 1930, 'Properties of soils which influence soil erosion,' *US Dept. Agric. Tech. Bull.,* 178.

MOLDENHAUER, W. C., WISCHMEIER, W. H. and PARKER, D. T., 1967, 'The influence of crop management on runoff, erosion and soil properties of a Marshall Silty Clay Loam.' *Soil Sci. Soc. Amer. Proc.,* 31, pp. 541–6.

NIR, D. and KLEIN, M., 1974, 'Gully erosion induced by changes in land use in a semi-arid terrain (Nahal Shiqma, Israel).' *Zeitschrift für Geomorphologi Supp Bd.,* 21, pp. 191–201.

RAPP, A., 1974, 'A review of desertization in Africa — water, vegetation and man.' *Secretariat for Int. Ecol., Sweden, Bull.,* 1.

RAPP, A., BERRY, L. AND TEMPLE, P., 1972, 'Soil erosion and sedimentation in Tanzania.' *Geografiska Annaler,* 54, pp. 105–9.

RAPP, A., LE HOUEROU, H. N. and LUNDHOLM, B., 1977, 'Can desert encroachment be stopped?' *Ecol. Bull.,* 24.

RAUSCHKOLB, R., 1971, 'Land degradation.' *Soils Bull.,* 13 (FAO, Rome).

SAVAT, J., 1977, 'The hydraulics of sheetflow on a smooth surface and the effect of simulated rainfall.' *Earth Surface Processes,* 2, pp. 125–40.

SMITH, R. M. and STAMEY, W. L., 1964, 'How to establish erosion tolerances.' *Jnl. Soil and Water Cons.,* 19, pp. 110–12.

VAN DOREN, C. A., STAUFFER, R. S. and KIDDER, E. H., 1950, 'Effect of contour farming on soil loss and runoff.' *Soil Sci. Soc. Amer. Proc.,* 15, pp. 413–17.

WISCHMEIER, W. H., 1959, 'A rainfall equation index for a universal soil loss equation.' *Soil Sci. Soc. Amer. Proc.,* 23, pp. 246–9.
1977, 'Soil erodibility by rainfall and runoff', *Erosion, Research Techniques Erodibility and Sediment Delivery,* ed. T. TOY (Geo Books, Norwich), pp. 45–56.

WISCHMEIER, W. H., SMITH, D. D. and UHLAND, R. E., 1958, 'Evaluation of factors in the soil loss equation.' *Agricultural Engineering,* 39.

YAKUBOV, T. F., 1969, 'Some results of wind erosion of soils.' *Pochvovedenize,* 12, pp. 115–27.

YOON, Y. N. and WENZEL, H. G., 1971, 'Mechanics of sheetflow under simulated rainfall.' *Proc. ASCE Jnl. Hyd. Div.,* HY9, pp. 1367–86.

Biosphere

13

Ecosystems and Communities: Patterns and Processes

CAROLYN M. HARRISON
University College London

13.1 INTRODUCTION

The mosaic of plant and animal assemblages forms the starting point of most biogeographical enquiries and this chapter presents a personal view of four major trends in research which have an important bearing on the types of explanations used to answer questions about how and why particular mosaics emerge.

First, the period since 1964 and the initiation of the International Biological Programme (IBP) has seen a great deal of research effort devoted to an understanding of the functioning of world ecosystems. During this programme ecologists, biologists and geographers have been encouraged to pool their knowledge and skills to provide rigorous descriptions and analyses of the world's forest, woodland, grassland, desert, and tundra ecosystems at a level of detail never previously attempted. Ecosystem studies, of course, have but a short tradition stemming from Lindeman's paper in the early forties on Cedar Lake (Lindeman, 1942) and followed by other well-known studies such as those of Teal (1957), and Odum (1957). These early contributions were of geographically discrete units such as lakes, springs and abandoned fields and few concerned themselves with a really extensive or ecologically complex ecosystems. The IBP attempted to deliberately rectify this deficiency by studying representative examples of the world's ecosystems ranging from the arctic tundra to the rainforest of the humid tropics. In doing so, the IBP spawned a host of new techniques for description and analysis and for the first time provided a standard basis for comparisons of the production potential of the biosphere, and a sound basis for assessing limits to production. It is to these latter studies that ecologically orientated efforts to improve food production must turn.

A second theme in biogeographical work concerns the study of plant succession. It is a theme which has a long tradition and inevitably generates controversy. Clements' (1916) early ideas promoted a somewhat misleading analogy between the development of a plant community over time and that of the life-cycle of an organism. More recent workers view succession as a process which is the product of plant-by-plant replacement and where the patterns generated by this replacement process have routine statistical properties. Whittaker's (1953) concept of the climax as a pattern of species abundances which, while locally constant, varies from place to place in a continuous fashion is now accepted by many biogeographers, but there is less agreement about certain common properties thought to be exhibited by successional or climax communities. In particular, the relationship between diversity and stability promoted by Odum (1969) has provoked much debate, and some authors such as Horn (1974) have gone so far as to suggest that the search for any universal and unifying theory of succession is a fanciful goal. Yet succession is an important concept for land management and conservation of resources, and the causes of argument themselves highlight deficiencies in our knowledge and understanding of biogeographical processes which need to be made good.

A third major impetus to biogeographical studies has come from plant and animal ecologists concerned with an understanding of the nature and causes of spatial variations of biotic communities on a more local scale. Whereas ecosystem studies have been couched in terms of the major biomes of the world, themselves defined by dominant life-forms and gross structural characters, community studies attempt to explain the considerable floristic and faunistic variations present within each of these major biome types. The work of Curtis (1959), Whittaker (1966) and others in the mixed, temperate forest zone illustrates how variation within, and between, forest types can be described and understood in terms of plant populations varying continuously in time and space. Their approach to vegetation studies based on gradient analysis emphasizes the important role of inter-specific competition and niche differentiation, and focuses attention on those resources which influence and limit particular plant populations. In this respect the gradient approach has moved away from a traditional classificatory approach with its emphasis on homogenous units of vegetation and reorientates enquiry to a consideration of directions or trends of variation and their underlying causes.

Finally, and fourthly, biogeography in America received an enormous fillip from the revolutionary work of MacArthur and Wilson (1967), who provided a theoretical basis for understanding how the biotic composition of different oceanic islands reflects a dynamic equilibrium between extinction and immigration processes. Their work and that of others such as Simberloff (1974), has demonstrated that the changing flora and fauna of different oceanic islands can be viewed as a compromise between processes of colonization on the one hand and the inevitable process of extinction on the other. That the establishment of an equilibrium population can be described in mathematical terms owes much to the tradition of population biologists, but the use of distance and area as independent variables explaining oberved differences between island faunas and floras complements the thinking of biogeographers such as Wallace (1892) and Willis (1922) who belonged to an earlier methodological tradition. Perhaps of greatest importance to applied studies however, is the realization that the processes at work on oceanic islands are also applicable to the culturally-induced island habitats of woods, hedges, ponds etc on land. In these latter cases the theory of island biogeography can be applied to wildlife conservation strategies and reserve design.

These four themes in research, namely studies of world ecosystems, ecosystem succession, community pattern and the biogeography of islands, form the basis of this chapter. Some will feel that other themes should have their place in the development of biogeography as a subject, but the purpose of this chapter is not to attempt a comprehensive review but rather to highlight some recent trends in research and to indicate how they relate to applied studies and to the influence of man. This is then succeeded in Chapter 14 by consideration of conservation of ecosystems.

13.2   ECOSYSTEM PATTERNS AND PROCESSES

The essence of all ecosystem studies is an elucidation of those complex interactions which weld together plants and animals and their physical environment into a single working whole. Fundamentally this means an understanding of energy and nutrient pathways, not only their identification but also an examination of their relative importance from one ecosystem to another, and their changing character through time. Such a complete understanding of these ecological processes would enable man ultimately to direct the functioning of natural ecosystems to his own needs. The achievements of ecosystem studies may have fallen short of these objectives but they have provided new insights into the natural complexities of different ecosystems and have pointed the way for further research.

First and foremost, many ecosystems studies have focused attention on food-webs and the pathways of solar energy and organic matter. Initially this has meant a preoccupation with the role of primary producers, the green plants of terrestrial ecosystems and phytoplankton and algae in the marine and freshwater ecosystems. The relative efficiencies of different plant assemblages in different environments for converting solar energy to a form which can be utilized by animals and man are now well known and the productivity of most of the major ecosystems of the world can be compared on a similar basis (see Fig. 13.1 after Leith and Whittaker, 1975). The global pattern of primary productivity that emerges lends support to earlier beliefs that there is a poleward decline in terrestrial productivity which is a general reflection of the solar energy environment, but also highlights 'hot spots' of productivity which impinge on this general trend and reflecting the interaction between vegetation, available moisture, and nutrient supplies. More noteworthy perhaps is the emerging pattern of marine production in which southern Antarctic waters are seen to be as productive as parts of the Atlantic Ocean which are warmed by the gulf-stream, and tropical oceans less productive than even the coldest deserts of the world. Such a pattern raises questions about supplementation of tropical food supplies from ocean fishing and clearly suggests that an answer lies firmly with improving the productivity of land-based agricultural systems.

The reasons for these major differences between terrestrial and marine ecosystems are concerned with the environmental limits to production and man's ability to reduce their influence in artificially cultivated systems. For example, the dependence of many land crop plants on supplements of nutrients or water can be made good by fertilizer

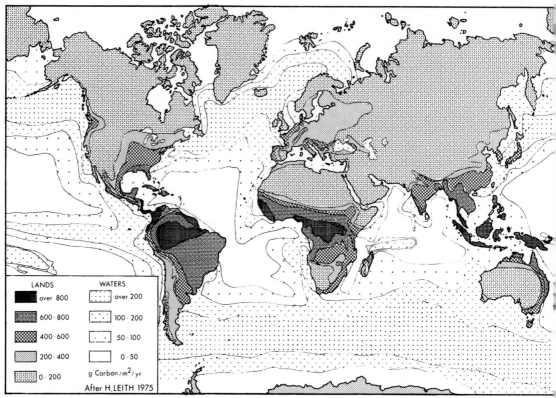

FIG. 13.1 PRODUCTIVITY PATTERN OF THE WORLD
(Source: Leith and Whittaker, 1975)

applications and irrigation, but in the sea, where nutrient supplies frequently limit production, artificial fertilizing only becomes a real possibility in lochs or estuaries, which are to some extent geographically and functionally contained units. Other ecosystem studies emphasise the diversity of productivities exhibited by similar vegetation types. For example Murphy (1975) demonstrates that the annual net primary production of tropical grasslands varies from as little as 180 g m^{-2} yr in Jodhpur in India for a dry year with 92·7 mm rain, to 650–3810 g m^{-2} yr for sites in Africa, Australia and India with annual rainfall figures of 700–1000 mm. The large range of primary productivities for a relatively small range of annual rainfall totals suggests that other variables such as rainfall periodicity, soil permeability and fertility, species character, and grazing pressure are important influences and that generalizations about the relationship between primary production and single environmental variables are notoriously inappropriate. That many ecosystem studies have not been able to demonstrate precisely how different environmental variables lead to changes in the productivities of an ecosystem is a reflection of the difficulties of observing and analysing multivariate and dynamic situations. Those studies which have attempted to allow for feedback operations among groups of variables have concerned predator/prey relationships either in the laboratory or using simplified food-chains and the monitoring of specific ecosystems subjected to environmental stress. In this latter category fall studies such as the irradiation of tropical rainforest in Puerto Rico, reported by Pigeon and Odum (1970) and the cutting over of a deciduous forested watershed in the Hubbard Brook (Bormann and Likens, 1970). The latter study perhaps comes closest to a more complete understanding of nutrient cycling in a forest ecosystem than any other ecosystem study. Basing their observations on a series of replicated watersheds, Bormann and Likens have been able to isolate specifically the role of decomposition in the terrestrial ecosystem and its crucial position in nutrient cycling. Where other ecosystem studies have been content to isolate the energy transfers between members of the above-ground grazing food chain, these workers have critically examined the very important function performed by organisms on and in the soil. Their results indicate that the capacity of an ecosystem to retain nutrients is dependent upon a finely balanced exchange between soils, plants, and the water entering the system. The loss of nutrients caused by cutting the forest is associated with the cessation of nutrient uptake by the plants, the larger quantities of drainage water passing through the system and to increased rates of decomposition resulting from such changes in the physical environment as higher soil temperature and soil moisture. In particular, they suggest that certain types of vegetation can inhibit nitrification in soils and so retard the production of highly leachable nitrate ions. This is not a new observation, but the demonstration that natural ecosystems are conservative of nutrients through the operation of relatively tight nutrient cycles points to the value of man's influence expressed in more conservative cultivation practices, in forestry and also to the value of traditional shifting cultivation practices in the humid tropics and elsewhere.

Perhaps it is in the field of pesticides that food-chain studies have proved most useful. Certainly the identification of the intricate food-webs of an estuarine ecosystem described by Woodwell *et al.* (1961) illustrates the numerous interrelated predator–prey links for a relatively small ecosystem, and it also demonstrates how a pesticide applied at the base of the food-web can be accumulated and concentrated in the tissues of higher trophic levels. The most thoroughly researched of all these food-chain links concerns the vulnerability of the top carnivores, especially birds of prey: for example, the decline in the population of the Peregrine Falcon, *Falco peregrinus*, both in Great Britain and North America, following upon the introduction of DDT and before its widespread use. The population decline was associated with reports of eggshell-thinning and failure of hatching and the suggested cause was DDT in the food supply. Recent work has proved beyond doubt that DDE, a metabolite of DDT, was present in these eggs and in amounts sufficient to cause thinning. The consequences of the widespread use of the broad-spectrum pesticides for wildlife and for crops and animals harvested by man seem clear. The higher up the food-chain an animal figures, then the more likely it is to accumulate high concentrations of a pollutant. Man's peculiar position as the organism which applies the toxin and as the top carnivore at the end of the food-chain, causes much concern and Rachel Carson's *Silent Spring* (1965) stands as an ominous warning to the unbridled use of non-specific and highly toxic substances. Subsequent studies have questioned the universality of the belief that organo-chloride insecticides such as DDT and other pollutants concentrate along food chains. For example Moriarty (1972) in reviewing several reported examples of this concentration effect, demonstrates that the amounts of chemicals retained by different organisms probably depends more on differing rates of metabolism and excretion. His work demonstrates clearly the need to link field observations with controlled laboratory studies, and to consider all of the routes along which a toxin could have entered an organism, not just in its food, and in particular to try and establish how long an organism has been exposed to a toxin and its ability to expel a toxin once absorbed.

13.3   ECOSYSTEM DYNAMICS THROUGH TIME

In the same way that it is tempting to provide universal laws about pesticide accumulation and trophic position, so too is it tempting to search for common properties of ecosystems as they change through time. Clements' early treatise on plant succession (1916) clearly adopted an integrated view of vegetation, soils, and topography, likening the plant community to an organism which grew and matured over time, ultimately arriving at a relatively stable stage, the climatic climax. During the first half of the twentieth century the Clementsian view of succession and the climax underwent considerable change. His views were challenged partly because he used climate as the dominant environmental factor which determined the composition of the climax community and partly because he adopted an organismic analogy to describe the way in which plants, animals and the environment interact. For example, Tansley (1935) suggested that several environmental variables other than climate were important in influencing the composition of a stable plant community and he proposed a polyclimax theory to replace Clements' monoclimax theory. Perhaps of more fundamental importance was the view expressed by Gleason (1926) and others, which emphasised the variability of species availability over time and the continuously varying nature of the environment through time. Gleason concluded that no plant assemblage could persist in a stable state over long time periods, rather it changed and fluctuated in response to frequently changing conditions. Such a view does not preclude repeated assemblages of plants from occurring for similar sets of environmental conditions, but it does suggest that this is likely to be a rare or chance event and that convergence of successions on a regional scale is highly improbable.

Many authors have since come to accept an interpretation of vegetational change through time in terms of fluctuating plant populations following Whittaker's (1953 and 1970) definition of the climax vegetation as a pattern of species abundances, which, while locally constant, varies from place to place in a continuous fashion. This definition is attractive and removes the need to prove that mutually beneficial co-operation between plants does exist or that integration in the plant community required by Clements and Tansley's organismic analogy can be demonstrated. But it does pose another question, that is how do we recognize whether a community is a climax community or not? It is here that the work of Margalef (1968) and Odum (1969) has relevance, for both authors attempt to provide a basis for characterizing successional and climax communities in terms of measurable properties. Of more significance, perhaps, is their attempt to review successional changes in terms of ecosystem processes rather than merely in terms of plant population dynamics. By doing so they illustrate how the functional processes of the ecosystem may be related to structural organization and how these relationships change over time. For example, Odum uses attributes relating to productivity, nutrient cycling, life-history and community structure, to present a model of successional trends and Margalef's ideas are largely in concert with this model.

Many authors have since provided empirical observations which would accord with these trends. For example Whittaker (1966) has shown that the selective removal of the oldest trees in climax woodland stands increased their overall productivity, and the Hubbard Brook work (Bormann and Likens, 1970) clearly illustrates the basis for closed nutrient cycling in late successional woodlands. Harper's work (1977) demonstrates that certain species illustrate adaptive strategies consistent with different positions along a successional sequence. Early successional species often produce large quantities of vagile offspring and late successional species produce fewer and larger offspring. By linking reproductive capability to environmental hazards, Harper recognizes r-species, those that spend most of their time in acts of colonization, and K-species which are specialists in resource-limited environments where there is intense competition from near neighbours. He demonstrates clearly however that r-species can be found in late successional communities, as long as environmental events occur which provide opportunities for colonization, and conversely notes that few K-species are associated with those environments which are characterized by frequent environmental disasters, as is the arctic tundra.

The more controversial aspects of Odum's model of ecosystem development concern the relations between diversity and stability and his concept of homeostasis. Diversity is usually used as a measure of the variety and abundance of different species. Early in succession a few species, some very abundant others rather rare, may comprise the community; in late succession very many more species of more even abundance are thought to be characteristic. Characterizing the diversity of communities is more readily undertaken for early successional stages than it is for the later stages, partly because it is one thing to record all plants in a sample but another to make an inventory of all the associated animals too. More significantly, however, as diversity increases locally, each species becomes rarer and hence rare species which contribute much to any index of diversity may be overlooked in a finite sample of the community. Diversity is thus an illusive property and rarely truly measurable. On a different basis Horn (1974) argues that in theory the diversity of the climax must be lower than that of some preceding stage, for if the climax is subjected to a disturbance that is small enough not to affect the relative proportions of the climax species, then a few additional pioneers are allowed to move in either as new species, or as colonists from existing populations rare in the undisturbed climax. Such an interpretation supports a view of the climax as a 'blurred successional patchwork' and Horn goes on to provide a mathematical Markovian model, which confirms that intermediate stages exhibit higher diversities than either early or late stages. Other workers have observed similar patterns (see Shafi and Yarranton, 1973), and recently Cole (1977) demonstrated that secondary forests attuned to former Indian burning practices and dominated by *Pseudotsuga menzii*, are now being replaced by a floristically impoverished *Abies grandis* community.

Stability of ecosystems is an equally illusive property to define and is best regarded as a complex notion involving both a resistence to perturbation and a constancy of composition over time. During succession, groups of species are replaced by new species combinations and this is conventionally how successional communities are defined. Logically therefore, stability as defined by frequency of population fluctuation, increases with succession. On the other hand, stability defined as the time taken for a community to return to its former condition, may actually decrease through time. For example, it takes much longer for the closed and varied vegetation of old established dunes to develop after heavy recreational pressure, than it does for the vegetation of marram dominated fore-dunes. Clearly the same argument would hold true if the disrupting agent was a storm surge rather than man's influence. The reasoning whereby Odum's strategy depicts the late stages of succession as stable in terms of a resistance to external perturbation (homeostatis), is based upon food-chain linkages, and itself is related to the work of MacArthur (1955). The initial assumption made is that a wide variety of food sources, and hence complex food-webs, decrease the frequency of explosions and crashes amongst animal and plant populations. In other words, a large number of alternative pathways that material (energy, organic matter, nutrients) can move along from the lowest to the highest tropic level buffers the community from external changes.

Such resilience to change however is not always a property of complex food chains, as is demonstrated by Watt (1968) and more comprehensively by May (1973). For example, Watt found that the stability of population fluctuations of boreal forest Lepidoptera pests was less, the larger the number of tree species fed on by any species of Lepidoptera; nevertheless the stability of populations was greater, the greater the number of competitor species for the tree species eaten by each species of Lepidoptera. He explains this paradox by identifying interspecific complexity on the one hand and trophic complexity on the other. In the case of the boreal forest insect pest, high interspecific complexity exists but with low trophic complexity because many species have wide-ranging diets and not restricted ones. The position of tropical forests can equally be interpreted as inherently unstable. May (1973) argues that the constant external environment of the humid tropics allows populations to evolve and adjust to a diversity of species interactions, which although individually efficient, are highly inefficient and vulnerable in a more variable environment, for example resulting from forest thinning or felling (Farnworth and Golley, 1974). Horn (1974) goes on to suggest that a constant environment allows a species to increase its efficiency by pursuing a narrower range of activities, resulting in a fragile diversity. Such 'fragile ecosystems' obviously would require protection from human disturbance and perhaps from well-intentioned management too.

Confronted with numerous and conflicting reports of 'simple-stable' and 'complex-unstable' ecosystems it is dangerous to seek a universal relationship between ecosystem diversity and stability however defined; nevertheless, an understanding of how the characters of different ecosystems vary in their persistence and resilience to change and the influence of human activity, is important to discussions of land management, conservation and the use of resources, as is discussed in Chapter 14.

13.4 COMMUNITY PATTERNS

A preoccupation with ecosystem dynamics through time tends to ignore a third dimension, namely that most geographic of all properties, spatial organization. The question of why vegetation varies from one place to the next, traditionally forms the starting point of numerous enquiries among geographers and ecologists and recently much published work has seen a convergence of opinion between what had previously been regarded as diametrically opposed schools of thought. Much of this controversy concerned the identification of what many plant sociologists saw as the basic unit of vegetation study, that is the plant association. Frequently linked with the work of European continental plant sociologists such as Braun-Blanquet (1932) and Tüxen and Ellenberg (1937) but also used widely by Clements and Tansley in North America and Britain, the term 'plant association' was often used to signify a discrete unit of vegetation defined by the possession of a definite floristic composition, uniform physiognomy, and occurrence in a uniform habitat. On the continent of Europe, Braun-Blanquet adopted a very restricted view of the association, based upon the possession of particular, character species and utilized it as the lowest unit in a classificatory hierarchy of plant communities. While the erection of each association was based on an abstraction derived from the comparison of numerous floristic tables drawn from related plant communities, the reality of the association as a unit which can be recognized in the field is also implicit in his method. It is this latter point which is perhaps most critical to Braun-Blanquet's view of the nature of vegetation variation. His method assumes that the vegetation of a region can best be described in terms of a number of spatially discrete plant associations. Such a view begs the question of a mechanism which promotes the development of distinct plant associations exhibiting the internal homogeneity required by the association concept. Certainly some species have similar though not identical ecological ranges, and where a site encompasses conditions suitable for each species, both could be expected to grow there in some form of equilibrium. But the inference of repeated associations in the landscape and a corresponding repetition of similar environments is that no two associations overlap and are separated by boundaries; variation in vegetation and environment is thus discontinuous. It is this latter view which other workers have challenged.

Prominent among the critics were Curtis (1959) and Whittaker (1966) following in the tradition set by Gleason (1926). In their opinion, plant communities are often less than discrete units and may be thought of as hardly more than the chance coming together of species whose tolerance ranges overlap. They do not invoke interdependence among species nor gestalt properties of vegetation where the whole is greater than the sum of the parts. Vegetation is then seen to vary continuously in space and through time. Expressed in these polemical terms, there appears to be little common ground between the two view points, but during the late 1960s and early 1970s many workers in America, Britain and Europe have provided a basis for reconciliation. Webb (1954, p. 364) for example says, not in exasperation but as a helpful observation, that 'the pattern of variation shown by the distribution of species among quadrats of the earth's surface chosen at random hovers in a tantalizing manner between the continuous and discontinuous'. He goes on to suggest that the only way in which categories that form a continuum may be delimited is by the erection of arbitrary boundaries. In essence therefore, classify if we must, but be aware of the arbitrary nature of the resulting classification, and make sure that the rules for defining units are clear. The belief that the method of plant sociology advocated by Braun-Blanquet was a craft which had to be learned, rather than a science, was founded partly on the failure of many English speaking workers to understand the 'rules' as applied and partly upon the changing nature of these rules as the method evolved in practice. John Moore's (1962) paper attempted to provide a clear, worked example of the different stages in the Braun-Blanquet method but his work and those of others at this time was largely overshadowed by the advent of many and varied multivariate classificatory techniques, whose rules were mechanistic and inviolable. By the late 1960s, the classification of plant communities had entered the realms of a multivariate science and the debate about integration in the plant association was subdued (Westhoff and van der Maarel, 1973). Multivariate methods were accepted as a means of arbitrarily defining abstract units of vegetation, where the units could be precisely defined by the possession of certain attributes or mathematical functions of similarity, dissimilarity, variation, etc. (Frenkel and Harrison, 1974).

In America at this time, several different techniques for the description and analysis of vegetational patterns were evolving, not least of them by Whittaker and his colleagues. Gradient analysis formed the basis for much of this work in which samples of vegetation are either arranged in relation to one or more axes which themselves may be assumed as given (direct gradient analysis) or are mathematically derived from a comparison of all samples (indirect gradient analysis) (Whittaker, 1973). Their work has revealed that it is possible to examine the inter-relationships between vegetation and environment in terms of environmental gradients, and that the structure and composition of vegetation can be accounted for by variation in a number of variables, such as moisture supply and nutrient availability, and by the effects of inter-specific competition. The plethora of different techniques available for analysis have been evaluated by a number of authors, for example Whittaker and Gouch (1973). One important conclusion from their studies is the isolation of the effects of the compositional difference of samples or Beta Diversity, from those of Alpha Diversity, that is species richness of samples. As the range of community diversity (β diversity) in a sample set increases, sophisticated indirect ordination procedures such as principal components analysis are less effective, while simple environmental gradient analyses such as the Wisconsin comparative ordination, with axes defined by most dissimilar samples, prove more useful. These two authors suggest that the complex mathematical treatments of vegetation such as PCA, factor analysis, and other related techniques should be regarded as speciality techniques likely to be of most use when results can be interpreted using detailed knowledge of species from other sources.

The uncompromising position adopted by earlier workers in vegetational science has today been eroded. Currently it is apparent that many European phytosociologists advocate and use an approach to the study of vegetational patterns in which tables of ecological groups of species are ordered in relation to one or more environmental gradient (Moore et al., 1970; Maarel, 1971). It is equally obvious that workers using a gradient approach to study find classificatory methods useful both at an initial exploratory phase and as a basis for summarizing findings. So that while the last ten years have seen a marked proliferation in the number of different computerised methods of analyzing vegetation patterns, the methodological rift which was apparent up to the mid 1960s has largely been healed and a new breed of 'vegetation ecologists' (Mueller-Dombois and Ellenberg, 1974) has emerged who are prepared to use both approaches to their studies, recognizing that in doing so, a more complete understanding and explanation can be achieved. This comprehensive approach is typified by the national survey of British vegetation being undertaken by the University of Lancaster and others for the Nature Conservancy Council (Rodwell, 1976). In this study both traditional phytosociological methods and multivariate analyses will be combined to provide a systematic and comprehensive, floristic inventory of British plant communities as they exist at the present time.

13.5 ISLAND BIOGEOGRAPHY

Over ten years ago MacArthur and Wilson (1967) revolutionized biogeography with the suggestion that the biota of any island is a dynamic equilibrium between immigration of species onto the island and extinction of species already present. In essence their theory of island biogeography suggested that species number would be constant over periods of several hundreds or thousands of years, while evolution would act gradually over geological time to increase the equilibrium number of species. Extinction, they argued should be viewed as a common, local, and ecologically important event and not as a rare, global, evolutionary one. Since first propounding this theory for oceanic islands, various workers have provided support for it from a variety of different environments and species groups, ranging from snails on Aegean islands (Hellier, 1976), to mammals on mountain tops (Brown, 1971), and duckweeds in inland lakes and ponds (Keddy, 1976).

Experimental work undertaken by Simberloff and Wilson (1969) first confirmed the dynamic equilibrium model (see Fig. 13.2). By removing all the fauna from a group of six, small red mangrove islands in Florida Bay and taking censuses of the animal populations over a period of years following, these authors confirmed that:

1 Near islands achieved an equilibrium more quickly than the most distant islands
2 The equilibria were dynamic in which species turnover, i.e. extinctions and new arrivals, occurred on a daily basis
3 Small islands have fewer species than large ones
4 Distant islands have fewer species than near islands.

These last two observations are consistent with the accumulated experience of very many natural historians, and MacArthur and Wilson formalized the relationship between the number of species S on an island and island area A in a double logarithmic relation:

$$S=S_o A^Z$$

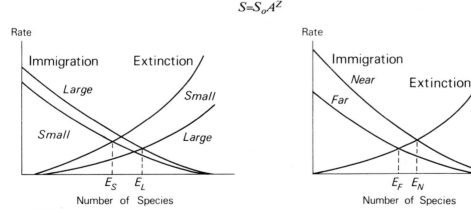

FIG. 13.2 COMPARISON OF ISLANDS
Large and small compared on left; far and near compared on right. (Source: MacArthur and Wilson, 1967)

Where S_o is a constant for a given species group in a given community, and Z usually assumes a value in the range 0·18–0·35. Numerous workers have confirmed the usefulness of this model for describing and predicting the relationship between an island's biota and island size. For example Diamond's (1973) study of the freshwater birds of the islands of the Bismarck Archipelago near New Guinea shows that the number of species gradually increases with increasing area, and that populations of similar sized islands decrease by a factor of 2 for each 2600 km distant from New Guinea, the possible source area (Fig. 13.3). This latter point illustrates the effect of increasing isolation on immigration rates.

The demonstration of dynamic equilibrium opened up new avenues of enquiry which involve population biologists, geneticists, and conservationists, as well as biogeographers. Diamond's work on the birds and plants of the Channell Islands off Southern California, provided clear evidence of species turnover in a fifty year period. He showed that the number of species of land and freshwater birds changed only slightly during this period, while composition changed markedly with a turnover of between 0·3 and 1·2 per cent per year. His work in the satellite islands of New Guinea approached equilibria in a different way. By using area, distance from New Guinea, and maximum elevation (an index of habitat diversity) as independent variables, Diamond calculated the equilibrium number of bird species for islands suspected of being out of equilibrium as a result of the severance of the land bridge connection with the mainland. Larger islands had estimated relaxation periods in the range of 7000 years, while the small island estimates are much lower. This type of analysis obviously has relevance to long term species conservation programmes on these islands and reinforces the observation that extinction should not be regarded as a rare event.

Extending his work further, Diamond (1975) has drawn analogies between the processes

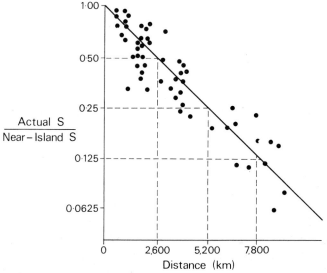

FIG. 13.3 RELATION BETWEEN NUMBER OF SPECIES AND DISTANCE OF ISLAND FROM THE COLONIZATION SOURCE IN AN ISLAND ARCHIPELAGO
(Source: J. M. Diamond, 1975)

operating on oceanic and land-bridge islands, and terrestrial islands separated from one another by inhospitable terrain. More pointedly, he has examined the implications of island biogeographic theory for conservation policies. Using the tropical rain forests as an example, he poses the question, what fraction of Amazonia must be left as rain forest to guarantee the survival of half of Amazonia's biota? Island biogeography suggests that the ultimate number of species that a reserve will save is likely to be an increasing function of the reserve's area. A rough rule of thumb suggests that a ten-fold increase in reserve area means a two-fold increase in the number of species, if we adopt the double logarithmic relationship advocated by MacArthur and Wilson. However, which species survive will depend upon the ecology of the site, and not just its geography, because it is evident that habitat diversity plays an important role in contributing to species survival. It is also clear that species in small reserves run the risk of likely extinction and that species in small reserves remote from others, run the risk of inevitable extinction. In addition, the character of the intervening land influences the effectiveness of dispersal and species immigration, and plant species producing heavy seeds and fruits are likely to be poor colonizers, as would certain non-volant animal groups. How long a species survives in a reserve thus depends upon this dynamic balance between immigration and extinction. The time taken to achieve a new equilibrium (relaxation time) in a newly created reserve, in a landscape which is subject to increasingly intensive agricultural usage varies with its size, its remoteness and the inhospitable nature of the dispersal barrier. Diamond suggests that a method of quantifying the survival prospects of a species is to determine its 'incidence function', that is, the fraction of 'islands' with a given total number of species (S) that a given species occurs on. In effect this means plotting the probability that a species will occur on an island of a particular size. By relating these incidence functions to the population density, reproductive strategy and dispersal ability of a species, it is possible to indicate likelihood of survival on a reserve of a given size. The predictive power of such theory is also borne out by Terborgh and Faaborg's (1973) work on bird populations of the West Indies.

The relevance of these island biogeographic studies to conservation policy is clearly demonstrated by Moore and Hooper's (1975) analysis of bird species numbers in British woods. Using presence and absence records of birds from some 433 woods, these authors conclude that the MacArthur and Wilson island model gives the best fit in a regression analysis of the species/area relation. They also suggest that the probability of a wood holding a particular species increases in linear relation to the logarithm of the area, although this relationship differs between species. For some species (e.g. robin, linnet, black cap, blue tit) there is a minimal area below which there is no likelihood of finding a nesting pair. On the other hand, other species such as blackbird, woodpigeon, dunnock, and chaffinch will nest in areas as small as 1 m². Their study suggests that most bird species are likely to occur in a wood of 100 ha or above, than in a smaller one, and that however many small copses a farmer retains or plants, these are unlikely to provide the species diversity of one large 100 ha wood. In an intensively farmed landscape therefore it should be the conservationists strategy to acquire and manage the largest reserve he can afford. That most woodland reserves in England and Wales are smaller than 100 ha poses very real questions about the continued survival of some of those species with small populations.

13.6   CONCLUSION

These four themes in biogeography are all likely to continue their importance in the future, but of the four, it is perhaps the last which poses and attempts to answer the most profound questions and those most pertinent to human influence. In Simberloff's view (Simberloff, 1974), the equilibrium hypothesis has proved useful in interpreting many and diverse insular situations, but more importantly, it has given biogeography general laws of both didactic and predictive powers which had previously been lacking. British biogeographers have been singularly slow to take up this approach and it is to be hoped that work such as that by Moore and Hooper (1975) in the field of wildlife conservation and that of Cody and Diamond (1975) in the field of evolutionary ecology heralds a period of research in which the implications of island biogeographic theory are fully explored.

REFERENCES

Bormann, F. H. and Likens, G. E., 1970, 'The nutrient cycles of an ecosystem.' *Sci. American*, 223, pp. 92–101.

Braun-Blanquet, J., 1932, *Plant Sociology*, trans., revised, and edited by G. D. Fuller and H. S. Conrad (McGraw-Hill, London).

Brown, J. H., 1971, 'Mammals on mountain tops: non equilibrium insular biogeography.' *Amer. Nat.*, 105, pp. 467–78.

Carson, R., 1965, *Silent Spring* (Penguin, Harmondsworth).

Clements, F. E., 1916, *Plant Succession. An Analysis of the Development of Vegetation* (Carnegie Inst., Washington), publ. 242.

Cody, M. L. and Diamond, J. M. (eds.), 1975, *Ecology and Evolution of Communities* (Belknap. Press, Harvard University, Cambridge, Mass.).

Cole, D., 1977, 'Ecosystem dynamics in the coniferous forest of the Willamette Valley, Oregon, USA.' *Jnl. Biogeog.*, 4, pp. 181–92.

Curtis, J. T., 1959, *The Vegetation of Wisconsin: An Ordination of Plant Communities* (Univ. Wisconsin, Madison).

Diamond, J. M., 1973, 'Distributional ecology of New Guinea birds.' *Science*, NY 179, pp. 759–69. 1975, 'The island dilemma: lessons of modern biogeographic studies for the design of natural reserves.' *Biol. Cons.* (7), pp. 129–46.

Farnworth, E. G. and Golley, F. B. (eds.), 1974, *Fragile Ecosystems: Evaluation of Research and Applications in the Neotropics*. A report of the Institute of Ecology (Springer Verlag., Berlin).

Frenkel, R. E. and Harrison, C. M., 1974, 'An assessment of the usefulness of phytosociological and numerical classificatory methods for the community biogeographer.' *Jnl. Biogeog.*, 1, pp. 27–56.

Gleason, H. A., 1926, 'The individualistic concept of the plant association.' *Bull. Torrey Bot. Club*, 53, pp. 7–26.

Harper, John L., 1977, *Population Biology of Plants* (Academic Press, London).

Hellier, J., 1976, 'The biogeography of Enid landsnails on the Aegean Islands.' *Jnl. Biogeog.*, 3, pp. 281–92.

Horn, H. S., 1974, 'The ecology of secondary succession.' *Ann. Rev. Ecol. and Systematics*, 5, pp. 25–38.

Keddy, P. A., 1976, 'Lakes as islands: the distributional ecology of two aquatic plants, *Lemna minor* and *L. taisulea*.' *Ecol.*, 57, pp. 353–9.

Leith, H. and Whittaker, R. H., 1975, 'Primary productivity of the biosphere.' *Ecol. Studies*, 14 (Springer Verlag, Berlin).

LINDEMAN, R. L., 1942, 'The trophic-dynamic aspect of ecology.' *Ecol.*, 23, pp. 399–418.

MAAREL, E. VAN DER, 1971, 'Basic problems and methods in phytosociology.' *Vegetatio.*, 22, pp. 275–83.

MacARTHUR, R. H., 1955, 'Fluctuations of animal populations, and a measure of community stability.' *Ecol.*, 36, pp. 533–36.

1972, *Geographical Ecology* (Harper and Row, London).

MacARTHUR, R. H. and WILSON, E. O., 1967, *The Theory of Island Biogeography* (Princeton University Press).

MARGALEF, R., 1968, *Perspectives in Ecological Theory* (Chicago Univ. Press).

MAY, R. M., 1973, *Stability and Complexity in Model Ecosystems* (Princeton Univ. Press).

MOORE, J. J., 1962, 'The Braun–Blanquet system: a reassessment.' *Jnl. Ecol.*, 50, pp. 701–9.

MOORE, J. J. *et al.*, 1970, 'A comparison and evaluation of some phytosociological techniques.' *Vegetatio.*, 20, pp. 1–20.

MOORE, N. W. and HOOPER, M. D., 1975, 'On the number of bird species in British woods.' *Biol. Cons.*, 8, pp. 239–50.

MORIARTY, F., 1972, 'Pollutants and foodchains.' *New Scientist*, 53 (787), pp. 594–6.

MUELLER-DOMBOIS, D. and ELLENBERG, H., 1974, *Aims and Methods of Vegetation Ecology* (J. Wiley, Chichester).

MURPHY, P. G., 1975, 'Net primary productivity in tropical terrestrial ecosystems,' *Primary Productivity of the Biosphere*, ed. H. LEITH and R. H. WHITTAKER (Springer-Verlag, New York).

ODUM, H. T., 1957, 'Trophic structure and productivity of Silver Springs, Florida.' *Ecol. Monog.*, 27, pp. 55–112.

1969, 'The strategy of ecosystem development.' *Science*, 164, pp. 262–70.

PIGEON, R. F. and ODUM, H. T., 1970, *A Tropical Rain Forest: a Study of Irradiation and Ecology at El Verde, Puerto Rico* (Washington, Atomic Energy Commission, Div. of Tec. Info.).

RODWELL, J., 1976, 'National vegetation classification.' *Bull. Brit. Ecol. Soc.*, VII, 2, pp. 6–7.

SHAFI, M. I. and YARRANTON, G. A., 1973, 'Diversity, floristic richness and species eveness during a secondary (post-fire) succession.' *Ecol.*, 54, pp. 897–902.

SIMBERLOFF, D. S., 1974, 'Equilibrium theory of island biogeography and ecology.' *Ann. Rev. Ecol. and Systematics*, 5, pp. 161–82.

SIMBERLOFF, D. S. and WILSON, E. O., 1969, 'Experimental zoogeography of islands. The colonization of empty islands.' *Ecol.*, 50, pp. 278–96.

TANSLEY, A. G., 1935, 'The use and abuse of vegetational concepts and terms.' *Ecol.*, 16, pp. 284–307.

TEAL, J. M., 1957, 'Community metabolism in a temperate cold spring.' *Ecol. Monog.*, 27, pp. 283–302.

TERBORGH, J. W. and FAABORG, J., 1973, 'Turnover and ecological release in avifauna of Mona Island, Puerto Rico.' *Auk*, 90, p. 759–79.

TÜXEN, R. and ELLENBERG, H., 1937, 'Der systematische und der ökologische Gruppenwert. Ein Beitrag zur Begriffsbildung und Methodik in der Pfanzensoziologie.' *Mitt. Florist. Soziol. Arbeitsgem*, 3, pp. 171–84.

WALLACE, A. F., 1892, *Island Life, or the Phenomena and Causes of Insular Faunas and Floras* (Macmillan, London).

WATT, K. F., 1968, *Ecology and Resource Management* (McGraw-Hill, London).

WEBB, D. A., 1954, 'Is the classification of plant communities either possible or desirable?' *Bot. Tiddskr.*, 51, pp. 362–70.

WESTHOFF, V. and VAN DER MAAREL, E., 1973, 'The Braun-Blanquet approach', *Ordination and Classification of Communities, Part V, Handbook of Vegetation Science*, ed. R. H. WHITTAKER (Junk, Hague).

WHITTAKER, R. H., 1953, 'A consideration of climax theory: the climax as a population and pattern.' *Ecol. Monog.*, 23, pp. 41–78.

1966, 'Forest dimensions and production in the Great Smoky Mountains.' *Ecol.*, 47, pp. 103–21.

1970, *Communities and Ecosystems* (Macmillan, New York).

1973, *Ordination and Classification of Communities. Part V, Handbook of Vegetation Science* (Junk, Hague).

WHITTAKER, R. H. and GOUCH, H. G., 1973, 'Evaluation of ordination techniques'. *Ordination and Classification of Communities*, ed. R. H. WHITTAKER (Junk, Hague), pp. 287–320.

WILLIS, J. C., 1922, *Age and Area: a Study in Geographical Distribution and Origin of Species*, (CUP, Cambridge).

WOODWELL, G. M. *et al.*, 1961, 'DDT residues in an East Coast estuary: a case of biological concentration of a persistent insecticide.' *Science*, 156, pp. 821–4.

14

The Conservation of Plants, Animals, and Ecosystems

I. G. SIMMONS

University of Bristol

14.1   A MOSAIC OF ECOLOGICAL SYSTEMS

If we accept that one of the aspirations of geography is the presentation of an holistic perspective which enfolds both objective measurements of the physical environment and man's perception and uses of it, then in the present context we might note that man's economy is inextricably linked with that of nature: *Oikos* is the root of both ecology and economics. The major purpose of this essay is to examine some ideas about the functioning and relationships of different types of ecological systems and to see how they might have relevance for our attitudes to land use and other spatial patterns, including 'conservation'.

As a starting point, I propose to take the basic concepts of Eugene Odum's (1969) paper, 'The strategy of ecosystem development', which are to some extent amplified in his book *Ecology* (2nd edn. 1975). These are based on the characteristics of the different stages which occur during succession in the development of ecological systems towards the self-maintaining equilibrium which is sometimes called the 'climax' or 'mature' condition. A number of characteristics of the energetics, community structure, life history, nutrient cycling, selection pressure and overall homeostasis change during succession and these are summarised in Table 14.1. The end point of successions, the mature stage, is especially notable for the high degree of internal symbiosis which is developed, for the closed nutrient cycles, and for the high degree of stability which is maintained, although piecemeal renewal is of course essential. In spatial terms, the trend is towards a mosaic of ecosystems moving towards maturity but with patches of early stages for example where new land is created, where catastrophic events have occurred, and where the death of large organisms such as senescent trees has taken place.

Table 14.1

A TABULAR MODEL OF ECOLOGICAL SUCCESSION: TRENDS TO BE EXPECTED IN THE DEVELOPMENT OF ECOSYSTEMS

Ecosystem Attributes	Developmental Stages	Mature Stages
Community Energetics		
1 Gross production/community respiration (P/R ratio)	Greater or less than 1	Approaches 1
2 Gross production/standing crop biomass (P/B ratio)	High	Low
3 Biomass supported/unit energy flow (B/E ratio)	Low	High
4 Net community production (yield)	High	Low
5 Food chains	Linear, predominantly grazing	Weblike, predominantly detritus
Community Structure		
6 Total organic matter	Small	Large
7 Inorganic nutrients	Extrabiotic	Intrabiotic
8 Species diversity—variety component	Low	High
9 Species diversity—equitability component	Low	High
10 Biochemical diversity	Low	High
11 Stratification and spatial heterogeneity (pattern diversity)	Poorly organized	Well-organized
Life History		
12 Niche specialization	Broad	Narrow
13 Size of organism	Small	Large
14 Life cycles	Short, simple	Long, complex
Nutrient Cycling		
15 Mineral cycles	Open	Closed
16 Nutrient exchange rate, between organisms and environment	Rapid	Slow
17 Role of detritus in nutrient regeneration	Unimportant	Important
Selection Pressure		
18 Production	Quantity	Quality
Overall Homeostasis		
19 Internal symbiosis	Undeveloped	Developed
20 Nutrient conservation	Poor	Good
21 Stability (resistance to external perturbations)	Poor	Good
22 Entropy	High	Low
23 Information	Low	High

Source: E. P. Odum (1969)

We can refine the classification by designating four main types of ecosystem:

1 Early succession systems with low biotic diversity and linear food chains and a high net community production
2 Mature systems which generally show the opposite characteristics from type (1). Notably, a lot of biomass is supported by the energy flow, relative to type (1), and the food chains are predominantly web-like, with the detritus stage very important
3 A mixture of (1) and (2), at times of general environmental change, for example, or where natural fires bring about the juxtaposition of the two types of system
4 Inert systems, with little or no life: volcanoes are one example, ice-caps another.

If we can place spatial boundaries to each of these types of system we have the basis for a compartment model which can link each to the other. In the natural world the tendency is for type 1 to become type 2 but within the transition, some reversion to type 1 is always happening. Similarly, type 1 systems encroach upon areas of type 4 when they colonize cold volcanic lava flows or recently deglaciated terrain, for example. Volcanic activity or glacier advance increases the area of type 4. There is no doubt that sometime during the present epoch, after the major post-glacial climatic adjustments but before both the intensive and extensive ecological impacts of agricultural societies, the world must have looked like a mosaic of those types. At present it clearly does not, but we are reminded by Odum that ecosystems manipulated by man can be described in the same terms as natural ecosystems. Thus early successional phases of natural ecosystems are equivalent to the simple systems of modern agriculture; the mixed systems of type 3 would apply to areas of mixed forest and farmland; the mature systems are the same in both, often being deliberately protected by man as parks or reserves; and the inert systems have their equivalents in cities (which like volcanoes give off SO_2 and particulate matter) and derelict land. Using the type and quantity of energy flow as a criterion, it is possible to combine both natural and man-manipulated ecological systems in one classification (Table 14.2). Here the source of the energy becomes a differentiating factor. Some systems (Type 1 in Table 14.2) are entirely solar-powered and may use only the incoming solar radiation which they fix as chemical energy. Others (Type 2) may receive a natural subsidy of organic matter brought in by natural processes, and estuaries are an obvious example of this type. Another major category of systems (Types 3 and 4) are those which are subsidized by man-procured supplies of fossil fuels (i.e. stored photosynthesis). Some systems are indeed subsidized by fossil fuels but remain fixers of solar energy (e.g. mechanized agriculture, short-cycle forestry); others, like the city, are entirely powered by fossil (and in some cases, nuclear) fuels. The absolute throughput of energy per unit area (energy intensity) increases through Types 1–4 (Table 14.2).

These ecosystem types can also be related to the compartment model. Successional systems belong to Type 3 and may have a very high energy intensity if powered by fossil fuel but little more than mature systems (which equate to Type 1) if only human energy is used upon them. There is no explicit equivalent of the mixed systems. The fuel-powered industrial systems (Type 4) are clearly the inert category and are parasitic on the others, although we should not underestimate either the organic production of city suburbs or the value of the information created and transmitted in urban structures.

The major inference we can make from the compartment model and from Table 14.2 is that human strategy runs counter to that of nature. Whereas nature moves towards mature, self-maintaining systems, man moves towards one type of successional phase at the expense of the mature and mixed systems, and increasingly requires subsidiary energy to keep up productivity. He also increases the area of inert systems. In turn, many kinds of wastes reduce the diversity of organisms present in all types of system (Woodwell, 1970). The further implication, not perhaps openly stated by Odum but clearly intended, is that the matrix of ecosystem types produced by the imposition of human strategies on those of nature will in future be subject to unpredictable and uncontrollable fluctuations, that is they will not be stable enough as a habitat for large numbers of our species.

The concept of stability in ecology is not, as is apparent from Chapter 13, an easy one. If it is used in two senses, (Hill, 1975), then the first relates to a change of species composition over time, and an early successional stage is obviously much less stable in this respect than a mature ecosystem. The second refers to the ability of an ecosystem to

Table 14.2
ECOSYSTEMS CLASSIFIED ACCORDING TO SOURCE AND LEVEL OF ENERGY

Ecosystem type	Annual Energy Flow Kilocalories per sq metre per year (Kcal/m²/yr)	
	Range	Average (Estimated)
1 Unsubsidized natural solar-powered ecosystems: e.g. open oceans, upland forests. Man's role: hunter–gatherer, shifting cultivation	1000–10000	2000
2 Naturally subsidized solar-powered ecosystems. e.g. tidal estuary, lowland forests, coral reef. Natural processes aid solar energy input: e.g. tides, waves bring in organic matter or do recycling of nutrients so most energy from sun goes into production of organic matter. These are the most productive natural ecosystems on the earth. Man's role: fisherman, hunter–gatherer	10000–50000	20000
3 Man-subsidized solar-powered ecosystems. Food and fibre producing ecosystems subsidized by human energy as in simple farming systems or by fossil fuel energy as in advanced mechanized farming systems: e.g. Green Revolution crops are bred to use not only solar energy but fossil energy as fertilizers, pesticides and often pumped water. Applies to some forms of aquaculture also.	10000–50000	20000
4 Fuel-powered urban–industrial systems. Fuel has replaced the sun as the most important source of immediate energy. These are the wealth-generating systems of the economy and also the generators of environmental contamination: in cities, suburbs and industrial areas. They are parasitic upon types 1–3 for life support (e.g. oxygen supply) and for food; possibly fuel also although this more likely comes from under the ground except in LDCs where wood is still an important domestic fuel.	100000–3000000	2000000

The most productive natural ecosystems and the most productive agriculture seem to have upper limits of c. 50000 Kcal/m²/yr.

Source: Odum (1975)

return to its original successional pathway or self-maintaining state after a perturbation and is perhaps of most interest here. Jowett (1972) makes the analogy of a marble in a saucer: if hit, the marble will normally return to its position in the depression but it can be flipped so hard that it rolls out of the saucer entirely and finds an equilibrium position elsewhere. Examples from human impacts on ecological systems are easy to find. Here we must note than in man-manipulated systems, large quantities of energy, somatic and fossil based, are spent in keeping marbles in saucers. The importance of stability to man is quite simple, because without it there can be no sensible prediction and hence no rational planning for the future, only an immediate single-purpose response to random events. The forms and scale that instability might take, resulting from unsuitable man-management of ecosystems are very varied and not easy to foresee in detail. Desertification, soil erosion, pest outbreaks, disease epidemics, famines, and climatic change are all among the possibilities.

Viewed in this context 'conservation' becomes an attitude towards stability on the continental and global scales in terms of ecosystem development. Those actions which tend to or seem likely to produce instability are seen as anti-conservative, whereas those which preserve or enhance stability are seen as part of the conservation effort. The general problem is to keep fluctuations of ecosystem parameters within manageable bounds, remembering the demands made by man upon the systems and the nature of the technology which he uses to manipulate them.

The bulk of this essay will be concerned with brief accounts of the main trends in the human use of the earth which tend to maintain or to diminish the overall stability of its ecosystems, both natural and managed. Readers will notice that few of the authors of the publications which are cited are geographers, because these are fields in which they have not generally been active, with the notable exception of desertification. In terms of a general synthesis by a geographer which strongly argues a particular case, a book by S. R. Eyre, *The Real Wealth of Nations* (1978) should be read.

14.2   ANTI-CONSERVATION TRENDS

14.2a   Intensification of agriculture

Without attempting a complete list of destabilizing influences, a number of outstanding degradational processes must be discussed, and leading the list are the effects of agriculture throughout the world. In many places the trend is towards intensification of production, which has several unfortunate results. The most obvious of these is the complete breakdown of the soil when it is tilled or cropped at an intensity which it cannot tolerate, the erosional results of which are discussed in Chapter 12. Such processes are now most common in less-developed countries (LDCs) where human population pressure is severely pressing upon food production, but similar trends can be observed in some developed countries (DCs) where, for example, the absence of ley phases in tight cropping schedules on land in areas of medium to high rainfall brings about dangerously low levels of organic matter and threatens the stability of the soils. The drainage of soils may be affected by plastic deformation and smear caused by wheels and tillage implements, and so poorer germination and restricted root development may lead to more open soil and hence more erosion (Agricultural Advisory Council, 1970). The extension of agricultural ecosystems into areas formerly consisting of mature systems also leads to immense losses of soil, including its constituent organic matter and mineral nutrients. Traditional systems of shifting agriculture usually replaced these losses during the fallow period, but the curtailment of the rotation cycle and the forcing of peasant agriculture onto very steep slopes, makes such recovery barely possible. The loss of mature forest on mountains then contributes to a higher frequency and height of floods lower down the river basin so that floods can often be seen as a symptom of intensified land use within a watershed (cf. Chapter 5). Even in a soil conservation-conscious nation like the USA, an average of 26·9 tonnes ha^{-1} of topsoil is lost annually from agricultural land. This material also reduces the usefulness of lakes, reservoirs and rivers to the tune of $500 million/yr, not counting the interference with the fish populations caused by enhanced silt loads and eutrophication. To offset the losses caused by such erosion, Pimentel *et al.* (1976) calculate that 494 000 kcal ha^{-1} (50 × 10^6 bbl oil/yr) of energy inputs have to be deployed on the land.

The biological components of agro-ecosystems can also be destabilized. In the past this was done mainly by weeds and pests, so that the application of a great battery of chemical agents has in general brought both ecological and economic stability to these systems. But, as is well known, the effect of some pesticides has been to bring about a deviation-amplifying condition. This has happened when an insect, for example, has built up resistance to a chemical biocide and has become a serious pest before new measures to combat it can be developed; or when the residues of a biocide have accumulated in a food chain and have proved lethal to a non-target organism which was an important structural member of a linked ecosystem. The breeding of plants for modern agriculture has meant a trend to genetic uniformity, in order to ensure an homogenous response to water and fertilizers, a single cropping time, and undifferentiated response to industrial processing. In the USA 53 per cent of the cotton crop is in three varieties, 95 per cent of the peanut crop in nine varieties, and 65 per cent of the rice crop in four varieties (NAS, 1972).

Such practices increase the vulnerability of crops to failure from events like drought and insect attack, since in a stand with genetic variety some individuals would probably survive. The problem seems especially worrying in LDCs where the Green Revolution has brought a considerable dependence upon high yield varieties with a large measure of genetic uniformity. Intensification of agriculture anywhere in the world brings problems arising from runoff containing fertilizers rich in nitrogen and phosphorous. This may often lead to nutrient enrichment of fresh waters with consequent disruption of plant and animal life, an algal 'bloom' often being the first sign. This process is called eutrophication, and it is also one result of the discharge of untreated sewage from towns and cities (Sawyer, 1966).

The outcome of certain human uses of the land, allied to possible changes in climate, is seen in the set of phenomena called 'desertification' (cf. Chapter 4). This is defined as 'the spread of desert conditions for whatever reasons, desert being land with sparse vegetation and very low productivity associated with aridity, the degradation being persistent or in extreme cases irreversible' (Grove, 1977). The conversion of land into a virtually inert system and its associated susceptibility to rapid erosion, clearly puts the process into the destabilizing category and while not perhaps the only causative factor, the effects of grazing, cultivation, woodland clearance, and borehole development are clearly not to be ignored (Mabbutt, 1977).

14.2b   Urbanization and industrialization

One major process of the present time which is common to most parts of the world is urbanization and industrialization, which usually produce inert systems in place of all the other categories, either directly because of their replacement by a built environment, or indirectly because the wastes reduce or even eliminate many forms of life. The quantity of land in urban–industrial use is very variable from nation to nation, even within industrial economies. In the USA, urban uses plus highways account for ca 29×10^6 ha of land (including 270000ha devoted to parking), which is 3·2 per cent of the total land area, including Alaska (Pimentel *et al.*, 1976). In England and Wales about 10 per cent was 'urban' in the 1960s (Anderson, 1977). However not all the urban–industrial land is built-up or inert and Best and Ward (1956) showed for example, that in a London suburb the value of food output unit area was close to that of the best farmland. Nevertheless, numerous examples of the more or less complete sterilization of land by dense building or by industrial heaps and holes, are easily brought to mind, with the presence of such wild life as starlings and pigeons being accounted for by the fact that they rely on energy imported from outside the city. Spin-off effects of cities and industries also destabilize ecosystems. Raw sewage and calefacted water, and acid mine drainage alike reduce the biological diversity and biomass of water bodies; fall-out of atmospheric contaminants may kill plants or reduce their growth rate; cumulative poisons like polychlorinated Biphenyls (PCBs) or methyl mercury may result in the mortality of many organisms including, in the latter case, man himself. Floods too, may be exacerbated by urbanization and its resultant rapid runoff (Chapter 5, Fig. 5.3). Not all such developments must necessarily be viewed as negative. Industrial life has so far brought a much more comfortable, and culturally enriched, life than subsistence agriculture or pastoral nomadism, and it can be argued that it is the surpluses generated by industrial life that permit the development and transmission of knowledge that will enable their destabilizing effects to be overcome. Again, the cities are indeed parasitic upon other systems for food and are obviously powered by non-renewable fossil fuels, but perhaps we burn these sources of energy in order to learn how to manage without them.

14.2c War

From being a localized phenomenon whose ecological effects were largely temporary, war has gradually become a destabilizer at virtually national scales with long lasting effects. Just as the shell holes of the first World War are preserved in a fossil form in Flanders, so much more freshly preserved are the immense numbers of craters which mark the Republic of Vietnam. The Vietnamese war of re-unification represents the first time that intentional anti-environmental actions were a major component of the strategy and tactics of one of the adversaries. Characterized by chemical warfare against mangroves, forests and crops, together with the effects of artillery on forests, with mechanized land clearing, and with attempts to wash out the Ho Chi Minh trail by artificial rain-making, industrialized war has perhaps never before been applied in such intensity to one country (SIPRI, 1976). Neither need it be forgotten that in the various forms of thermonuclear device, the potential for ecological destabilization and even sterilization is very high (SIPRI, 1977).

14.2d Species extinction

Many of the conversions of system type discussed above produce one kind of result in the extinction of species of plants and animals either by direct extirpation or by alterations of habitats. This is no new event for it has been postulated that during the terminal Pleistocene of many parts of the globe, Palaeolithic hunters administered a *coup-de-grâce* to many genera and species of large mammal herbivores which were at the time under stress because of environmental change; this is the phenomenon known as 'Pleistocene overkill' (Martin, 1973). But the expansion of empires in the seventeenth century and, above all, the coming of the industrial revolution have increased the numbers of animal extinctions to a considerable degree. Between 1600 and 1699, there were twelve extinctions of mammals, birds and marsupials, whereas from 1900 onwards there have been at least eighty-one (Ziswiler, 1967). The extinction rate of species and subspecies appears to be about one per year, compared with one every ten years from 1600 to 1950, and perhaps one every 1000 years during the period of the great dying of the dinosaurs (Myers, 1976). In spatial terms, the vulnerability of oceanic islands is outstanding, and the recently occupied continents of North America and Australia also contribute relatively high numbers to the total.

The extinction of plants must not be forgotten. Habitat changes have ousted populations of many plants and, as with animals, island floras have been particularly susceptible to introduced herbivores like goats and sheep. For example, Phillip Island, 1610 km east of Sydney, was thickly forested in 1774 but within 100 years grazing animals had helped to remove the forest and two endemic plants along with it. The extinction of the dodo, endemic to Mauritius, seems to have meant that a particular tree (*Calvaria major*) has not regenerated since the late seventeenth century. The hypothesis is that crushing in the dodo's gizzard was necessary to break the tough endocarp and allow the excreted but undigested seed to germinate. Now there are only about thirteen trees left, all over 300 years old (Temple, 1977). Globally, one estimate (Tinker, 1971) suggests that 20 000 plant species, representing about 10 per cent of the world's flora, are now threatened with extinction. Among such plants there are probably many that might be useful as well as beautiful or stabilizing components of ecosystems.

14.2e Response to fluctuations

The destabilizing effects of human activity affect the world mosaic of ecosystems via several environments. They may interact adversely with the soil and produce soil erosion or infertility; they may enter the system via the role of organisms in terms of extinctions of taxa, or via the introductions of exotic species which can 'explode' into new habitats, or via the direct and residual effects of pesticides. Organisms can also be affected by wastes. The hydrological cycle can be destabilized by particular land-use practices, producing floods, and we are ignorant of the possible future fluctuations of ecosystems which may result from the extension of inert systems and the wastes they produce. Such processes are aptly described by Woodwell (1970), who brings together the evidence from deforestation, toxins, radioactive wastes, eutrophication, and other similar processes. He talks in terms of such events as the loss of biotic diversity; of population instability, especially of small rapidly reproducing organisms such as rodents and insects; and of a movement towards a world which is not self-augmentive and homeostatic any more but requires instant tinkering to patch it up. Each act of tinkering, however, generates a need for further action by man.

The diverse suite of actions by human societies to manipulate ecological systems can be summed up in terms of the use of energy. Increasing amounts of energy (which at present mostly come from fossil fuels in the DCs) are being demanded in order to cope not only with the requirements for intensification and industrialization described above but with the aftermath of destabilized systems. Because fossil fuels are finite in amount, the alternative energy futures of man hold the key to the type of attitude we develop towards the world's mosaic of ecological systems: this is discussed further in a later section (14.3e) of this essay.

14.3 CONSERVATION TRENDS

14.3a Protected landscapes and ecosystems

We recall that ecological stability seems to be more assured if there is at any time a mosaic of seral and mature communities, and it is common in many nations to give protected status to portions of the latter. So we have landscapes and ecosystems which have a special legal status and are called by a particular name: National Parks, Nature Reserves, Game Parks, and Protected Landscapes are all part of this complex (Simmons, 1974). Not all are mature systems, but those which are may in fact be relict in character since a great deal of their surroundings will have been converted by human activity to successional or inert systems. The formally protected lands are complemented by residual areas of 'unused' land where little manipulation of the ecosystems has taken place, e.g. parts of the Sahara and the North American tundra. The present network of parks and resources excluding Greenland and Antarctica, covers about $1 \times 10^6 km^2$ or 1·1 per cent of the earth's land surface, and concentrates on the spectacular and unique rather than the representative. Of the world's 198 terrestrial biotic provinces, more than one-quarter contain no protected areas and a further 15 per cent, only one (Myers, 1976).

At one end of a continuous spectrum of protected ecosystems are the reserves for the perpetuation of a single species of plant or animal. Such taxa are generally rare or threatened, or possess some symbolic value. Examples are the reserves in Hokkaido for the sacred Japanese crane, the refuge for the California Condor (probably down to 40 individuals), and the reserves protecting rare orchids on the Chalklands of southern England. More complex is the protection of habitats together with their characteristic assemblages of plants and animals. These may be formed into linked systems as with the wildfowl refuges along the migratory flyways of North America, or the wetlands incorporated in the international convention on Wetlands of International Importance. Relict forests in agricultural zones of the world are often in this category, as with the remnants of the Scots Pine forests of Scotland, or the last fragment of beech-spruce urwald in southern Bohemia. The class may also extend to underwater areas, where coral reefs or other biotically rich habitats form the nuclei of marine parks as in the Caribbean, Japan, Hawaii, and Australia (Greden, 1975; Ray, 1976).

The size and configuration of both these two types of reserve is usually determined as much by politics as biology but a spatially isolated population is also genetically isolated. If rates of dispersion across unprotected terrain are low, it may be essential to keep corridors of protected land connecting the main reserves. It is possible to design reserve systems whose shape and spatial relations maximise the probability of survival of a threatened species (Fig. 14.1).

A last category contains the reserves which are large enough to protect whole sets of ecosystems. Nations with a lot of wild terrain can designate large zones, usually called National Parks (NP), for the conservation of biota and scenery. The parks of eastern and central Africa are perhaps the best known (Myers, 1972a, 1972b), apart from the

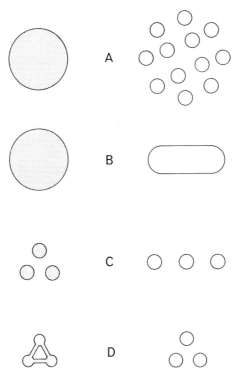

FIG. 14.1 THE GEOMETRICAL RULES OF DESIGN OF NATURAL PRESERVES, BASED ON CURRENT BIOGEOGRAPHIC THEORY
The design on the left results in each case in a lower spontaneous extinction rate than the complementary one on the right. Both the left and the right figures have the same total area and represent preserves in a homogeneous environment. (A) a continuous preserve is better than a fragmented one, because of the distance and area effects. (B) a round design is best because of the peninsula effect. (C) clumped fragments are better than those arranged linearly, because of the distance effect. (D) if the preserve must be divided, extinction will be lower when the fragments can be connected by corridors of natural habitat, no matter how thin the corridors. (Source: Wilson and Willis, 1975)

spectacular landscapes of some of the western Cordillera of North America at places like Banff–Jasper NP, Greater Lake NP, and Yosemite NP. Even arid areas have their attractions: Australia has an Ayers Rock–Mount Olga National Heritage Area (Lacey and Sallaway, 1975). The great national parks grade into the concept of the wilderness, where large areas of land are set apart from all economic use and man's intrusion is allowed at only a low level for scientific purposes or low-impact recreations such as hiking. Wilderness zones may be designated within national parks or, as in the USA, special wilderness areas may be created within reserves of public land (Simmons, 1966). The outstanding instance is Antarctica where both the ecosystems and the ice-cap are protected by the Agreed Measures appended to the Antarctic Treaty of 1959. A wide measure of protection is given to native flora and fauna and steps are taken to prevent the introduction of alien species which might 'explode' in the biologically impoverished ecosystems of the edges of Antarctica. Whether such protection will persist in the face of increased economic interest in the resources of the Antarctic seas, its continental shelf and sub-ice rocks, and more tourism, will be a severe test of international co-operation in conservation.

14.3b Outdoor recreation

Protection of landscapes and ecosystems is often carried out because areas of wild terrain are particularly sought after for recreation, whether of an active kind such as walking and camping or more inactive such as looking at scenery or cultural monuments (Simmons, 1975). Thus landscapes may be protected which also contain wild biota or even manipulated ecosystems as islands of less intensively used land within more productive landscapes. The national parks of England and Wales, and Japan, are of this type. However, management primarily for recreation may be antithetical to conservation aims (Hendee and Stankey, 1973). In any event, large numbers of visitors usually lead to erosion around gathering places and paths, and to disturbance of the fauna. Scavengers will inevitably emerge from the fauna, so that bighorn sheep, British hill sheep, blue jay, yellow hammer, and brown bear have all been observed by the author filling this niche. During 1963–9, conflicts between people and grizzly bears in Yellowstone National Park had come to the level of 4·4 injury incidents per year (Cole, 1974). Nevertheless, multiple-purpose management can, if sophisticated, often keep people away from the sensitive ecosystems or biota and so the recreation areas can contribute to the total of protected systems.

14.3c Conservation of genetic variety

The tendency of modern resource development is to reduce biological diversity both by eliminating species and by breeding domesticated animals and plants from a very narrow range of genotypes within favoured taxa. The practice of cloning, which gives a uniform genetic inheritance to a crop and is therefore economically desirable, makes the organism an easy prey to climatic shifts, the introduction of new pests or indeed the resurgence of old ones. So the buildup of resources of genetic diversity becomes an important part of any conservation programme (Frankel and Hawkes, 1975; Rendel, 1975).

One place where this is done is within zoological and botanical gardens where the biota are kept in the live adult form and encouraged to breed, surplus young being sold off or possibly stored as seed in the case of plants. Apart from preserving threatened species and providing a pool of genetic material, it is argued that the presence of such places is a part of public education which will encourage the formation of biological awareness in individuals. Some zoos and gardens try harder than others in this aim. Outside such institutions, particular individuals and corporate bodies may be involved in special breeding and protection programmes. At one extreme is an idiosyncratic individual who breeds varieties of cows, pigs, or chickens which are ignored by the commercial producers and hence decline in numbers; at the other is the international research institute determined to collect and preserve viable seeds from every known variety and ancestor of a particular crop plant, for the whole world.

Gene banks may therefore take a variety of forms. Not to be underestimated are the wild places of the earth, small and large, though the latter are clearly the most important. Nobody knows when a wild plant or animal may come in useful, whether for introducing a new strain into an already domesticated crop, or as an agent of biological control, or as an entirely new crop. To take a single example, a desert plant of South-western North America, the jojoba (*Simmondsia chinensis*) may provide the first acceptable substitute for sperm whale oil, as well as the basis of an economy for native peoples of the region (Yermanos, 1974; Maugh, 1977). The preservation of variety in agriculture also seems useful, as well as under experimental conditions at research institutes. These places may also have the facilities and skills to preserve parts of plants and animals containing genetic information, for example as viable seeds, cell and tissue cultures, as frozen semen, or perhaps even frozen embryos in the future.

14.3d Low-impact technology

The fundamental idea behind low-impact (or 'appropriate', 'alternative', 'soft') technology is access to a modest sufficiency of resources without destabilizing ecological systems. It often represents an attempt to live within the carrying capacity of existing systems rather than replace them with new systems requiring a higher energy subsidy to keep them stable. The ideas seem to have a particular relevance for LDCs, where the shortcomings of technology transfer from the DCs have in recent years gained a lot of exposure, not only in ecological terms (Taghi Farvar and Milton, 1972) but also socially, particularly in terms of rural unemployment and the drift to the cities. Thus 'ecodevelopment', relying more on equilibrium sources of energy together with local and traditional methods of self-sufficiency, and possibly the industrialization of renewable resources such as plant life, is the basis for exchange economies.

In terms of food, for example, the universalist approach of the Green Revolution is modified by a renewed valuation of traditional agriculture which must have had a survival value, since it lasted so long. Local knowledge and practical science is reflected in efficient and stable systems like shifting agriculture, wet rice terraces, and Polynesian gardens. There are many difficulties, such as how do the systems cope with rapid population growth; and can an ecologically sound way of using Amazonia be developed before it is all destroyed? The potential of forests can also be revalued since they can be sources of industrial materials, animal fodder, and human food; forest cultivation then becomes a more progressive idea than forest clearance. Protein from leaves and 'weeds' (both terrestrial, and aquatic, such as water hyacinth) might add to food resources. The development of energy sources for LDCs would aim at freeing them from dependence on DC technology and prices for all except perhaps a highly technical finishing stage to an otherwise decentralized, labour-intensive process. But in rural areas, solar collection, wind power, and fermentation of organic wastes may all be applicable, and preferable to centralized sources of power (Sachs, 1976). Such an approach has another long-term advantage: it promotes a diversity of cultural–economic ways of life, which in total has greater survival value than a reduction to a uniformity of lifestyle. It seems probable, too, that the greater social security brought about by modest but significant developments may also change attitudes to the frequency of child-bearing, and so induce slower rates of population growth.

In developed countries, the scope for a change to low-impact technology might at first sight seem limited. Yet, spurred on by such eventualities as soil erosion, pesticide problems, and the spiralling price of energy, new approaches in agriculture are being brought forward. In intensive cultivation, for example, the practice of minimum tillage, a return to crop rotation, and the use of animal manure are advocated (Pimentel *et al.*, 1973) as ways of reducing both soil erosion and fossil energy input. Techniques to use lower quantities of inputs more effectively, such as fertiliser balls inserted near growing roots and trickle irrigation from punctured pipes, are also becoming more popular. Beyond these measures there are the ideas of 'radical agriculture' which transforms the consumers' demands: more self-sufficiency even in urban areas, less demand for a standardized product of a quasi-industrial nature, fewer convenience foods and, above all, less consumption of meat, underlie this approach (Merrill, 1976; Wardle, 1977).

Both DCs and LDCs share the continuing interest in biological control of weeds and pests rather than the indiscrimate use of chemicals. True, new generations of toxins are appearing which are more target-selective and less persistent than for example the chlorinated hydrocarbons much used in the 1950s and 1960s, but where effective integrated bioenvironmental control (using techniques such as parasites, pathogens and predators of pests, habitat manipulation, crop spacing and rotation, breeding of host plant resistance, and sterile insect methods) can be maintained or introduced, it may often be cheaper and certainly pose fewer environmental problems such as the growth of resistant strains and the accumulation of poisonous residues (Huffaker, 1971).

14.3e Soft energy paths

Interest is growing in both DCs and LDCs in alternatives to the obvious development of nuclear fission power (especially the plutonium-fuelled fast breeder reactor) as a partial or total replacement for fossil fuels as they become too expensive to extract. The advantages of alternatives are seen as both environmental and social. The former includes the avoidance of the waste storage problems associated with the breeder reactor, and the large quantities of waste heat per unit of electricity generated, and the latter the turning away from possibilities of home-made plutonium bombs and the creation of a small priestly elite who would virtually control the country. Alternative energy sources, like geothermal and tidal power would, it is argued, be decentralized, would not add to the heat burden of the atmosphere, and would not pose waste-disposal difficulties. For DCs, the problem seems to be whether, even with energy conservation programmes of a high order, such alternative sources could, even with fossil fuels still available, provide sufficient power to obviate the breakdown of a basically industrial society. The Public Inquiry in 1977 into the proposals to expand the nuclear fuel reprocessing plant at Windscale, Cumbria, England, highlighted many of these questions and there is a lively debate between the protagonists of nuclear power (e.g. Hoyle, 1977) and those enthusiastic about alternative paths (e.g. Lovins, 1977). The events of 1979 have given added point to these arguments.

14.4 ALTERNATIVE FUTURES

14.4a Introduction

Implicit in most of this essay has been the idea that different kinds of man–nature relationships are possible in the future. In the various alternatives, all the different ecological systems so far discussed might have modified roles, along lines which are not yet clear.

The essay ends, then, with an outline of some of the emerging characterization of two of the most discussed of alternative man–nature relationships of the future. Some possible influences of these upon the ecosystem types of the front section of this chapter are given in Table 14.3.

14.4b   A technological future

This set of ideas is committed to the continued growth of material economies as measured by such indices as Gross National Product. The champions argue that not only will DCs continue to prosper, but LDCs will rise out of their poverty given such growth. The basis of the development will be the application of science and technology in the manipulation of nature, in which the provision of abundant industrial energy will be the key factor. Nuclear sources, initially fission reactors and eventually fusion sources, will provide this. Another key development may well be the ability to engineer the very basis of genetic material by recombining DNA molecules within the cell to produce a predictable genetic product. For example, non-leguminous plants might have nitrogen-fixing properties 'programmed' into appropriate cells. This technique has its obvious dangers, however, like nuclear power, and there is a similar controversy over its development and application.

The power to change nature conferred by possession of such tools might be used in two ways. The first would be to replace natural systems with totally man-made ones: given enough energy, we speculate that food can be made from granite, and artificial environments for all purposes, including the storage or processing of wastes, would replace all natural systems except presumably the atmosphere. At the extreme we can imagine with Fremlin (1964) the whole globe being covered with a building 2000 stories high and housing 60000×10^{12} people (890 years' growth from the present at 2 per cent per annum). The second alternative in this category would be the uncoupling of the ecosphere and the econosphere by, for example, putting industrial plant underground or offshore, and possibly enclosing cities under domes. If food could be made industrially, given cheap energy, then many ecosystems could be allowed to revert to a natural or mature condition. An initial step towards such a condition is described by Häfele (1974) who envisages power production from nuclear parks of about 30 GW capacity producing electricity and hydrogen (to replace hydrocarbon fuels) and containing all stages of the plutonium economy on one site. Waste heat would be initially led away to areas of the ocean which were thought to be ecologically and meteorologically insensitive to the release of the waste heat.

The outer limits of such a future system are perhaps dependent upon the dispersal of heat to space without disrupting the predictable functioning of the atmosphere. At present this is not generally a problem, although some cities do produce large quantities of heat. The Hung Hom district of Hong Kong produces a midwinter flux which is twice the daily incoming solar radiation in midwinter, and in Sydney, Australia, the equivalent figure in 49 per cent. Cities might therefore be the first places to have a limit imposed upon their growth by problems of waste-heat dispersal (Jaske, 1973; Kalma and Newcombe, 1976).

Table 14.3

POSSIBLE TRENDS OF ECOSYSTEM TYPES IN DIFFERENT MAN–NATURE RELATIONSHIPS

Ecosystem type	Man–nature relationship		
	Equilibrium economy	High technology (coupled)	High technology (uncoupled)
Successional (natural)	Retain place in natural and near-natural ecosystems.	Enlarged number in places where other systems disturbed by high energy impact, e.g. derelict land; energy affluence means reclamation of them not thought necessary.	Retain place in natural and near-natural systems.
Successional (man–directed, e.g. agricultural)	Important but subject to limits.	Agriculture presumably of much diminished importance (except? for luxury prod.) .'. of indust. prodn. of food. Systems may revert to mature or be made inert.	If food prodn. industrialized, then more natural area and role of these ecosysts much diminished; undergo succession to more mature system type.
Mature	Important as stabilizing element in total system, also for recreation, aesthetic purposes etc.	Diminished role if wholesale manipulation of ecosystems made feasible by abundant energy: converted to inert or successional systems.	Enhanced area if pressure to convert to inert or successional states are lifted. Importance is highly valued.
Inert (natural)	Balance achieved: nature creates but succession converts to productive systems. Major perturbations (e.g. ice-ages) may disrupt 'normal' balance.	Likely to diminish if adapted to inert but man–directed systems.	As in equilibrium economy: 'normal' place in nature.
Inert (man–directed)	Compartment is restricted in area and where possible diminished by reclamation/conversion to productive systems.	Likely to increase as technology and/or built environment substitutes for all kinds of organic systems.	Specialized areas of inert systems may act as a decoupling buffer between ecosphere and econosphere.
Mixed systems	Highly valued as diverse environments with capability of transforming successional mature.	Conversion to inert or purely mature systems likely.	Conversion to mature systems likely.

14.4c   An equilibrium future

This alternative view is based on a rejection of the idea of the replacement of nature and the substitution of a desire to live within the constraints of the ecological envelope. Thus the soft energy paths and appropriate technology discussed above are seen as a mode of life for DCs and LDCs alike. The throwaway society is replaced by one devoted to recycling and thrift, whether it be food, energy, or materials. In such a view, natural areas (i.e. mature systems) would lose any taint of being luxuries for the rich and be seen as essential parts of the environmental tesselation (Odum and Odum, 1972).

Transition to such an economy for the DCs might well be difficult (Daly, 1973; Pirages, 1977) but it is seen as having greater survival value, and greater cultural and ecological diversity than the alternative (Dasmann, 1975). One other immediate problem would be whether it could cope with population growth in certain key LDCs in Asia and Latin America. In this economy, traditional economic growth would not be sought, partly because the role of centralized authority necessary for such a trajectory would be greatly diminished. Neither would there be outer limits in the sense used: perception of the limits of the environment would feed back into a desire to live well within them, i.e. at a preferred level of resource use rather than an absolute level, rather like many hunter–gatherer peoples of the past (Lee and De Vore, 1968).

14.5 FINAL WORDS

It is clear that 'conservation' is intimately bound up with both economics and ecology. Both disciplines have in the past spoken with different languages and it is clear that they need to be brought together for a rational assessment of the futures available to the economics of both man and nature. One approach is that of Georgescu-Roegen (1971, 1976) who uses the concept of entropy (defined loosely as a measure of the unavailable energy in a thermodynamic system) as a linkage between the two conceptual systems. His argument summarized suggests that economic activity increases entropy and that only life processes have a relatively stable and low-entropy state. The stock of low entropy on the globe is limited and solar energy alone represents a virtually free low-entropy addition. This is all perhaps a complicated way of saying that he favours an equilibrium man–nature relationship and in this he is joined by many others, for it has the additional advantage of not foreclosing any options for our descendants.

But the technological alternative has its attractions, not the least being a promise of material plenty for all. The responsibility for deciding which path is taken is decided by societies through their political institutions or through the transformation of the consciousness of individuals. The discipline of geography, in its desire to understand the connections of the natural and social systems of the world, could well have a key role to play in promoting an enhanced awareness, among a wider range of people, of their linkages with the earth.

REFERENCES

AGRICULTURAL ADVISORY COUNCIL (UK), 1970, *Modern Farming and the Soil*, (HMSO, London).
ANDERSON, M. A., 1977, 'A comparison of figures for the land use structure of England and Wales in the 1960s.' *Area*, 9, pp. 43–5.
BEST, R. and WARD, D., 1956, *The Garden Controversy*. Wye College Papers in Agricultural Economics.
COLE, G. F., 1974, 'Management involving grizzly bears and humans in Yellowstone National Park 1970–3.' *BioScience*, 24, pp. 335–8.
DALY, H. E. (ed.), 1973, *Toward a Steady-State Economy* (Freeman, San Francisco).
DASMANN, R. F., 1975, *The Conservation Alternative* (J. Wiley, Chichester).
EYRE, S. R., 1978, *The Real Wealth of Nations* (Edward Arnold, London).

FRANKEL, O. H. and HAWKES, J. G., (eds.) 1975, *Crop Genetic Resources for Today and Tomorrow*. IBP Studies, 2 (CUP, Cambridge).

FREMLIN, J. H., 1964, 'How many people can the world support?' *New Scientist*, 24, pp. 285–7.

GEORGESCU-ROEGEN, N., 1971, *The Entropy Law and the Economic Problem* (MIT Press, Cambridge, Mass.).

1976, *Energy and Economic Myths* (Pergamon Press, Oxford).

GREDEN, G. A., 1975, 'Managing marine national parks: conflicts in reserve exploitation.' *Proc. Ecol. Soc. Austral.*, 8, pp. 147–55.

GROVE, A. T., 1977, 'Desertification.' *Progress in Physical Geog.*, 1, pp. 296–310.

HÄFELE, W., 1974, 'A systems approach to energy.' *American Scientist*, 62, pp. 438–47.

HENDEE, J. A. and STANKEY, G. H., 1973, 'Biocentricity in wilderness management.' *Bioscience*, 23, pp. 535–8.

HILL, A. R., 1975, 'Ecosystem stability in relation to stresses caused by human activities.' *Can. Geog.*, 19, pp. 206–20.

HOYLE, F., 1977, *Energy or Extinction?* (Heinemann, London).

HUFFAKER, C. B. (ed.), 1971, *Biological Control* (Plenum Press, NY).

JASKE, R. T., 1973, 'An evaluation of energy growth and use trends as a potential upper limit in metropolitan development.' *The Science of the Total Environment*, 2, pp. 45–60.

JOWETT, D., 1972, 'The quantitative assessment of environmental impacts.' *Environmental Impact Analysis: Philosophy and Methods*, (ed. R. B. DITTON and T. L. GOODALE (University of Wisconsin, Madison), pp. 127–36.

KALMA, J. D. and NEWCOMBE, K. J., 1976, 'Energy use in two large cities: a comparison of Hong Kong and Sydney, Australia.' *Env. Studs.*, 9, pp. 53–64.

LACEY, J. A. and SALLAWAY, M. M., 1975, 'Some aspects of the formulation of a planning policy for the management of the Ayers Rock–Mount Olga National Heritage Area.' *Proc. Ecol. Soc. Austral.*, 9, pp. 256–66.

LEE, R. B. and DE VORE, I. (eds.), 1968, *Man the Hunter* (Aldine Press, Chicago).

LOVINS, A. B., 1977, *Soft Energy Paths: Towards a Durable Peace* (Penguin, Harmondsworth).

MABBUTT, J. A., 1977, 'Climatic and ecological effects of desertification.' *Nature and Resources*, 13, pp. 3–9.

MARTIN, R. S., 1973, 'The discovery of America.' *Science*, 179, pp. 969–74.

MAUGH, T. H., 1977, 'Guayule and jojoba: agriculture in semi-arid regions.' *Science*, 196, pp. 1189–90.

MERRILL, R. (ed.), 1976, *Radical Agriculture* (Harper, NY).

MYERS, N., 1972a, 'National parks in savannah Africa.' *Science*, 178, pp. 1255–63.

1972b, *The Long African Day* (Macmillan, NY).

1976, 'An expanded approach to the problem of disappearing species.' *Science*, 193, pp. 198–202.

NAS (NATIONAL ACADEMY OF SCIENCES, USA), 1972, *Genetic Vulnerability of Major Crops* (NAS, Washington, D.C.).

ODUM, E. P., 1969, 'The strategy of ecosystem development.' *Science*, 164, pp. 262–70.

1975, *Ecology*, (Holt, Rinehart and Winston, NY, 2nd edn.).

ODUM, E. P. and ODUM, H. T., 1972, 'Natural areas as necessary components of man's total environment.' *Trans. 37th North American Wildlife and Natural Resources Conf.*, pp. 178–89.

PIMENTEL, D. *et al.*, 1973, 'Food production and the energy crisis.' *Science*, 182, pp. 443–9.

PIMENTEL, D. *et al.*, 1976, 'Land degradation: effects on food and energy.' *Science*, 194, pp. 149–55.

PIRAGES, D. C. (ed.), 1977, *The Sustainable Society* (Praeger, NY).

RAY, G. C., 1976, 'Critical marine habitats.' *Proc. Int. Conf. on Marine Parks and Reserves, Tokyo 1975* (IUCN Pubs. N.S., 37, Morges, Switzerland).

RENDEL, J., 1975, 'The utilization and conservation of the world's animal genetic resources.' *Agriculture and Environment*, 2, pp. 101–19.

SACHS, I., 1976, 'Environment and styles of development', *Outer Limits and Human Needs*, (ed. W. H. MATTHEWS (Dag Hammarskjold Foundation, Uppsala), pp. 41–65.

SAWYER, C. N., 1966, 'Basic concepts of eutrophication.' *Jnl. Wat. Poll. Control Fedn.*, 42, pp. 737–44.

SIMMONS, I. G., 1966, 'Wilderness in the mid-twentieth century USA'. *Town Planning Rev.*, 36, pp. 249–56.

1974, *The Ecology of Natural Resources* (Edward Arnold, London).

1975, *Rural Recreation in the Industrial World* (Edward Arnold, London).

SIPRI (STOCKHOLM INTERNATIONAL PEACE RESEARCH INSTITUTE), 1976, *Ecological Consequences of the Second Indo-China War* (Almqvist and Wiksell, Stockholm).

SIPRI, 1977, *Weapons of Mass Destruction and the Environment* (Taylor and Francis, London).

TAGHI FARVAR, M. and MILTON, J. P. (eds.), 1972, *The Careless Technology: Ecology and International Development* (Natural History Press, NY).

TEMPLE, S. A., 1977, 'Plant-animal mutualism: co-evolution with Dodo leads to near extinction of plant.' *Science*, 197, pp. 885–6.

TERBORGH, J., 1975, 'Faunal equilibria and the design of wildlife preserves', *Tropical Ecological Systems Trends in Terrestrial and Aquatic Research*, ed. F. B. GOLLEY and E. MEDINA, Ecological Studies, 41 (Springer-Verlag, NY), pp 369-80.

TINKER, J., 1971, 'One flower in ten faces extinction.' *New Scientist*, 50, pp. 408–13.

VAN DEN BOSCH, R. and MESSENGER, P. S., 1973, *Biological Control* (Intertext Books, NY).

WARDLE, C., 1977, *Changing Food Habits in the UK* (Earth Resources Research, London).

WILSON, E. O. and WILLIS, E. O., 1975, 'Applied biogeography', *Ecology and Evolution of Communities*, ed. M. L. CODY and J. M. DIAMOND (Bel Knapp, Cambridge, Mass.), pp. 522–34.

WOODWELL, G. M., 1970, 'Effects of pollution on the structure and physiology of ecosystems.' *Science*, 168, pp. 429–33.

YERMANOS, D. M., 1974, 'Agronomic survey of jojoba in California.' *Econ. Bot.*, 28, pp. 160–74.

ZISWILER, V., 1967, *Extinct and Vanishing Animals* (Springer-Verlag, NY).

Conclusion

15

A Perspective

K. J. GREGORY and D. E. WALLING

The preceding fourteen chapters have highlighted the significance of human impact on a variety of physical landscape processes. Coverage has been selective rather than exhaustive and it must be recognized that the writers have been primarily concerned with processes which have traditionally attracted the interest of the physical geographer. A study of human impact within a broader environmental context would require extension to embrace such topics as air and water pollution, the fate of pesticides and other contaminants, changes in global biogeochemical cycles, landscape aesthetics and the general quality of life. Within the confines of this review it has been emphasized that man must be viewed as a highly significant force in landscape processes and as a potent instrument of change within the contemporary global environment. However, one must not lose sight of the fact that human impact is relevant to only a minute recent portion of the long evolution of this planet and that the earth has been subjected to many other natural processes of great significance in the past. These include continental drift, tectonic activity, changing sea levels, the appearance of living organisms, Pleistocene climatic change, and other climatic fluctuations. To many, therefore, it might seem that man's activity merely concerns the embroidery on a mantle which has been tailored through the long period of geological time.

Whatever its relative significance, it is clear that human impact must play an increasingly important role in influencing environmental processes and that this impact must be accepted by the physical geographer as an important influence on contemporary landscape dynamics. Many of the changes occasioned by man's activities are also detrimental in the broader environmental context and the physical geographer must be aware of the wider implicatioins of this theme in the field of environmental management. For example, increased flood magnitude consequent upon forest clearance and land-use change clearly demonstrates the impact of man on hydrological and channel processes, but increased flood stages may in turn cause havoc and devastation where river flood plains are occupied by intensive human activity. Similarly, it has been pointed out in Chapter 4 that desertification with its many attendant problems of human starvation and migration may be the result of albedo changes associated with overgrazing, soil erosion, and increases in the dust content of the atmosphere. Situations whereby the intensification of human activity, related to increased population and advancing technology, gives rise to mounting environmental problems possess many of the features of negative feedback and have stimulated an increased awareness of the interactions between man and his environment. Attention is now being given to these problems at the international level (15·1); procedures are being developed to assess the potential impact on the environment of proposed developments in order to avoid future problems (15·2); new strategies are being adopted to monitor changes in the global environment (15·3); and these are beginning to witness the use of models (15·4) in order to test the likely impact of alternative development scenarios and to point towards the strategy with the least environmental cost. These themes usefully demonstrate the wider relevance and implications of man's impact on physical landscape processes.

15.1   INTERNATIONAL AWARENESS

The growth of international awareness and activity in the field of man-environment interactions has been marked in the literature by the appearance of a considerable number of acronyms referring to the organisations and projects involved. A rapid enumeration suggests that the number involved may approach thirty or more and whilst most readers will be familiar with such better known examples as MAB, UNEP and SCOPE, others such as GEMS, GARP and SCOR may be less well known.

Both the International Biological Programme (IBP) (1964–74) conceived by the International Council of Scientific Unions (ICSU), and the International Hydrological Decade (IHD) (1965–74) and its successor the International Hydrological Programme (IHP) sponsored by UNESCO, have devoted a considerable portion of their activities to the study of the impact of human activity on hydrological and ecological processes. The Man and the Biosphere (MAB) programme initiated by UNESCO in 1970 was, however, explicitly devoted to the theme of management problems arising from interaction between human activities and the natural system and the work of this intergovernmental body has been admirably complemented by the interdisciplinary scientific co-operation fostered by the Scientific Committee on Problems of the Environment (SCOPE). SCOPE was created in 1969 by ICSU as a means of co-ordinating the relevant activities of its constituent organisations and has as its purpose 'advancing knowledge of the influence of human activities upon the environment and the effects of the resulting changes on human health and welfare, with particular attention to those influences which are global or common to several nations.' The United Nations Conference on the Human Environment which was held in Stockholm 1972 must also rank as an important landmark in the development of international awareness. This focused public interest on the problems of environmental pollution and gave prominence to the problems of the planning and management of human settlements in the context of environmental quality. The associated United Nations Environmental Programme (UNEP) is specifically concerned with implementing a number of the recommendations of the Stockholm Conference and the United Nations Conference on Desertification which was held in Nairobi in 1977 arose from this body.

Other international bodies which have sponsored investigations on man's impact on the physical environment include the Scientific Committee on Oceanic Research (SCOR) which has directed attention to the pollution of the Baltic and to more general aspects of river inputs to ocean systems (the RIOS project). In addition is the Scientific Committee on Water Research (COWAR) which was responsible for the organization of the conference and workshops held in Alexandria in 1976 on the problems of arid land irrigation (Worthington, 1977) and for the interdisciplinary international conference on man-made lakes which took place in Knoxville in 1971 (Ackermann *et al.*, 1971). The International Geographical Union (IGU) has also contributed to this activity and its Commission on Man and the Environment has produced reports dealing with the environmental effects of technological developments (Nelson and Scace, 1974) and of complex river development (White, 1977).

Since 1970, the terms 'environment', 'pollution', and 'environmental quality' have been used very emotively in certain contexts. The recent increase in international awareness and activity relating to man's impact on the environment must therefore be viewed partly as the result of increasing scientific understanding of the potential problems and partly as a response to popular emotions. Together these two forces are beginning to create a background which should afford new insights into the nature and wider implications of this impact and an improved appreciation of means of regulating future human activity to reduce this impact.

15.2 MONITORING PROGRAMMES

One outcome of the growing awareness of the significance of human impact on landscape processes has been the application of modern technology to monitoring this impact, particularly by the development of national and international monitoring programmes. Recent advances in the use of satellite imagery offer many potential applications in this field, for example in sensing thermal pollution of water bodies, and short-term changes in vegetation cover and surface albedo (cf. Chapter 4). The use of automatic data collection platforms linked to satellite data retransmission systems similarly offers great potential for real-time data collection, particularly from remote areas.

The development of monitoring programmes designed specifically to monitor human impact has been justified as a means of establishing a baseline against which to evaluate future changes; of providing a synoptic view of contemporary environmental conditions; of detecting current changes; and of affording a means of predicting future trends. The network of benchmark catchments established in the USA may be viewed in the light of these aims. In this case, the objective is not to monitor change but to identify a baseline of minimum human interference against which data assembled from other stations subject to human activity may be evaluated (Cobb and Biesecker, 1971). Similarly the Harmonized Monitoring Programme operated by the Department of the Environment in the UK aims to provide a countrywide survey of water quality trends in major rivers and an evaluation of the flux of material from the land into the oceans.

At the international level, UNEP have established an International Referral System for Sources of Environmental Data (IRS) but problems of data consistency and comparability and of uniformity of objectives will clearly hinder its value. These problems underlie proposals for a unified global environmental monitoring programme and the concept of a Global Environmental Monitoring System (GEMS) was endorsed by the United Nations Conference on the Human Environment in Stockholm in 1972. A number of strategies have been formulated to further this objective (e.g. SCOPE, 1973; National Academy of Sciences, 1976; Rockfeller Foundation, 1977), but its full implementation faces several difficulties. Thus whilst its value is voiced with enthusiasm, many scientific, technological and procedural problems must be overcome. In some instances current technology cannot provide the instrumentation necessary to document parameters implicit in such a scheme. Developments in global atmospheric monitoring would, however, seem to have successfully overcome many of these practical difficulties and the Global Atmospheric Research Programme (GARP) organized jointly by ICSU and the World Meteorological Organization (WMO) and established in 1967 has made considerable progress in its attempt to assemble a global data set adequate to provide initial and verifying data for global models of atmospheric behaviour. On a more limited scale the study of long-range transport of air pollutants undertaken in Europe by the Organization for Economic Cooperation and Development (OECD) warrants mention. Work is currently in progress at the Monitoring and Assessment Centre (MARC) set up under the auspices of SCOPE and of UNEP at Chelsea College, University of London, to pursue the objective of a fully operational GEMS programme and future developments must be awaited with cautious optimism.

15.3   ENVIRONMENTAL IMPACT ASSESSMENT

Recognition of the many detrimental effects of human activity on environmental processes points to the need to assess the potential impact of future activity and to regulate or modify this impact accordingly. The terms environmental impact assessment (EIA) and environmental impact statement (EIS) are now widely used to refer to studies and statements which firstly attempt to produce estimates of future environmental changes attributable to a proposed action, and secondly attempt to suggest the likely impact of these changes on man's future well-being. In the United States, the National Environmental Policy Act (NEPA) of 1969 marked an important turning point in this context, because the Act required Federal Agencies to produce environmental impact assessments for all major actions, or in legislative terminology to 'identify and develop methods and procedures which will ensure that presently unqualified environmental amenities and values are given appropriate consideration in decision-making along with economic and technical considerations.' By the end of 1975 statements on nearly 7000 actions had been filed with the United States Council on Environmental Control. Subsequently a number of American states and cities have adopted similar requirements and the EIA system has become firmly established as a means of providing a measure of environmental protection. This situation is having widespread repercussions in many other countries and the governments of West Germany, France, Denmark, and Eire have enacted similar legislation.

An environmental-impact statement commonly requires discussion of the physical, biological, and social conditions existing prior to the proposed action; of the expected environmental impacts on these existing conditions; of any adverse effects which cannot be avoided should the proposal be implemented; of viable alternatives to the proposed action; of the relationship between local short-term uses of the environment and the maintenance of long-term productivity and stability; and of any irreversible and irretrievable commitments of resources which would be involved in the proposed action. It is clear that any attempt to 'cost' environmental changes and damage associated with a proposed development and to thereby evaluate alternative management or development strategies will prove difficult and at best *highly* subjective. Furthermore, discussion of impact in terms of human well-being and health, amenity, recreation, agricultural productivity, and similar criteria is beyond the scope of this text, but the EIA system must be viewed as a mechanism whereby many of the effects of man on landscape processes discussed in the preceding chapters can be given formal consideration.

A number of different methodologies and procedures have been developed for formulating an environmental-impact statement (e.g. SCOPE, 1975) and that produced by Dr Luna Leopold and his colleagues of the United States Geological Survey (Leopold *et al.*, 1971) may be briefly reviewed. The Leopold system involves an open-cell matrix listing 100 project actions along the horizontal axis and 88 environmental characteristics and conditions which might be affected by those actions displayed along the vertical axis. These are listed in Table 15.1. A ranking system scaled from 1 (least severe) to 10 (most severe) is used to designate the *magnitude* and *importance* of each possible impact. The system has been criticized because of its bias towards the physical–biological environment, but neither this limitation nor the details of the ranking system need concern us here. Table 15.1 may be viewed as a synthesis of much of the material contained in this book and as a means of ensuring that its implications are not neglected when large-scale developments are planned.

Table 15.2 illustrates the reduced matrix presented by Leopold *et al.* (1971) to demonstrate the most significant facets of an environmental-impact assessment for a proposed phosphate mining development in the Los Padres National Forest, California. Clearly, the success of this type of analysis depends on the extent and adequacy of knowledge concerning the interaction of environmental processes and attributes with human activity. It must be accepted that in certain cases this knowledge will be lacking and that improved understanding of the effects of man on physical landscape processes will in turn improve the capability for producing meaningful environmental impact assessments. Many of the environmental problems associated with the development of the Alaskan oilfields might have been foreseen and reduced if a fuller understanding of the dynamics of permafrost areas (cf. Chapter 9) had been available. As knowledge accumulates, environmental-impact statements will no longer be restricted to *identification* of the major areas of impact and a general measure of their severity but will provide *quantitative* estimates of the extent of the changes involved. An essential prerequisite for such progress is the development of process-based modelling strategies which allow the effects of particular human activities to be predicted and the impact of a number of alternative development scenarios to be simulated.

Table 15.1

THE LEOPOLD MATRIX FOR ENVIRONMENTAL IMPACT ASSESSMENT

Part A Environmental 'characteristics' and 'conditions' (vertically in the matrix)	Part B Project actions (horizontally in the matrix)

A PHYSICAL AND CHEMICAL CHARACTERISTICS

1 EARTH
a) Mineral resources
b) Construction material
c) Soils
d) Land form
e) Force fields & background radiation
f) Unique physical features

2 WATER
a) Surface
b) Ocean
c) Underground
d) Quality
e) Temperature
f) Recharge
g) Snow, ice & permafrost

3 ATMOSPHERE
a) Quality (gases, particulates)
b) Climate (micro, macro)
c) Temperature

4 PROCESSES
a) Floods
b) Erosion
c) Deposition (sedimentation, precipitation)
d) Solution
e) Sorption (ion exchange, complexing)
f) Compaction and settling
g) Stability (slides, slumps)
h) Stress-strain (earthquake)
i) Air movements

B BIOLOGICAL CONDITIONS

1 FLORA
a) Trees
b) Shrubs
c) Grass
d) Crops
e) Microflora
f) Aquatic plants
g) Endangered species
h) Barriers
i) Corridors

2 FAUNA
a) Birds
b) Land animals including reptiles
c) Fish & shellfish
d) Benthic organisms
e) Insects
f) Microfauna
g) Endangered species
h) Barriers
i) Corridors

A MODIFICATION OF REGIME
a) Exotic flora or fauna introduction
b) Biological controls
c) Modification of habitat
d) Alteration of ground cover
e) Alteration of ground-water hydrology
f) Alteration of drainage
g) River control and flow modification
h) Canalization
i) Irrigation
j) Weather modification
k) Burning
l) Surface or paving
m) Noise and vibration

B LAND TRANSFORMATION AND CONSTRUCTION
a) Urbanization
b) Industrial sites and buildings
c) Airports
d) Highways and bridges
e) Roads and trails
f) Railroads
g) Cables and lifts
h) Transmission lines, pipelines and corridors
i) Barriers including fencing
j) Channel dredging and straightening
k) Channel revetments
l) Canals
m) Dams and impoundments
n) Piers, seawalls, marinas & sea terminals
o) Offshore structures
p) Recreational structures
q) Blasting and drilling
r) Cut and fill
s) Tunnels and underground structures

C RESOURCE EXTRACTION
a) Blasting and drilling
b) Surface excavation
c) Subsurface excavation and retorting
d) Well drilling and fluid removal
e) Dredging
f) Clear cutting and other lumbering
g) Commercial fishing and hunting

D PROCESSING
a) Farming
b) Ranching and grazing
c) Feed lots
d) Dairying
e) Energy generation
f) Mineral processing
g) Metallurgical industry
h) Chemical industry
i) Textile industry
j) Automobile and aircraft
k) Oil refining
l) Food
m) Lumbering

Table 15.1—*continued*

Part A Environmental 'characteristics' and 'conditions' (vertically in the matrix)	Part B Project actions (horizontally in the matrix)

Part A Environmental 'characteristics' and 'conditions' (vertically in the matrix)

C **CULTURAL FACTORS**
1 LAND USE
a) Wilderness & open spaces
b) Wetlands
c) Forestry
d) Grazing
e) Agriculture
f) Residential
g) Commercial
h) Industrial
i) Mining & quarrying

2 RECREATION
a) Hunting
b) Fishing
c) Boating
d) Swimming
e) Camping & Hiking
f) Picnicking
g) Resorts

3 AESTHETICS & HUMAN INTEREST
a) Scenic views and vistas
b) Wilderness qualities
c) Open space qualities
d) Landscape design
e) Unique physical features
f) Parks & reserves
g) Monuments
h) Rare & unique species or ecosystems
i) Historical or archaeological sites and objects
j) Presence of misfits

4 CULTURAL STATUS
a) Cultural patterns (life style)
b) Health and safety
c) Employment
d) Population density

5 MAN–MADE FACILITIES AND ACTIVITIES
a) Structures
b) Transportation network (movement, access)
c) Utility networks
d) Waste disposal
e) Barriers
f) Corridors

D **ECOLOGICAL RELATIONSHIPS SUCH AS:**
a) Salinization of water resources
b) Eutrophication
c) Disease-insect vectors
d) Food chains
e) Salinization of surficial material
f) Brush encroachment
g) Other

OTHERS

Part B Project actions (horizontally in the matrix)

n) Pulp and paper
o) Product storage

E **LAND ALTERATION**
a) Erosion control and terracing
b) Mine sealing and waste control
c) Strip mining rehabilitation
d) Landscaping
e) Harbor dredging
f) Marsh fill and drainage

F **RESOURCE RENEWAL**
a) Reforestation
b) Wildlife stocking and management
c) Ground water recharge
d) Fertilization application
e) Waste recycling

G **CHANGES IN TRAFFIC**
a) Railway
b) Automobile
c) Trucking
d) Shipping
e) Aircraft
f) River and canal traffic
g) Pleasure boating
h) Trails
i) Cables and lifts
j) Communication
k) Pipeline

H **WASTE EMPLACEMENT AND TREATMENT**
a) Ocean dumping
b) Landfill
c) Emplacement of tailings, spoil and overburden
d) Underground storage
e) Junk disposal
f) Oil well flooding
g) Deep well emplacement
h) Cooling water discharge
i) Municipal waste discharge including spray irrigation
j) Liquid effluent discharge
k) Stabilization and oxidation ponds
l) Septic tanks, commercial & domestic
m) Stack and exhaust emission
n) Spent lubricants

I **CHEMICAL TREATMENT**
a) Fertilization
b) Chemical deicing of highways, etc.
c) Chemical stabilization of soil
d) Weed control
e) Insect control (pesticides)

J **ACCIDENTS**
a) Explosions
b) Spills and leaks
c) Operational failure

OTHERS

Source: Leopold *et al.* (1971)

15.4 MODELLING STRATEGIES

A variety of modelling strategies have been employed in investigating and assessing the impact of man on environmental processes, and a simple three part classification could distinguish conceptual, statistical, and simulation models. Conceptual models may be

Table 15.2
THE REDUCED DATA MATRIX FOR A PHOSPHATE MINING ENVIRONMENTAL IMPACT STATEMENT

Project Actions	II B.b. Industrial sites and buildings	II B.d. Highways and bridges	II B.h. Transmission lines	II C.a. Blasting and drilling	II C.b. Surface excavation	II D.f. Mineral processing	II G.c. Trucking	II H.c. Emplacement of tailings	II J.b. Spills and leaks
Environmental 'characteristics' and 'conditions'									
I A.2.d. Water quality					2/2	1/1		2/2	1/4
I A.3.a. Atmospheric quality						2/3			
I A.4.b. Erosion	2/2				1/1			2/2	
I A.4.c. Deposition, Sedimentation	2/2				2/2			2/2	
I B.1.b. Shrubs					1/1				
I B.1.c. Grasses					1/1				
I B.1.f. Aquatic Plants					2/2			2/3	1/4
I B.2.c. Fish					2/2			2/2	1/4
I C.2.e. Camping and hiking					2/4				
I C.3.a. Scenic views and vistas	2/3	2/1	2/3		3/3		2/1	3/3	
I C.3.b. Wilderness qualities	4/4	4/4	2/2	1/1	3/3	2/5	3/5	3/5	
I C.3.h. Rare and unique species	2/5			5/10	2/4	5/10	5/10		
I C.4.b. Health and safety								3/3	

Source: Leopold *et al.* (1971)

viewed as the least sophisticated mathematically. They involve little more than the recognition of important process interrelationships and causal mechanisms, and rely heavily upon the experience and intuition of the personnel involved. Thus a forester may have developed a detailed appreciation of the links between forest clearance, logging practice, runoff processes, erosion, nutrient cycling, and stream response and may be able to predict the impact of a particular forest-management programme on river water quality. Similarly a civil engineer or soil scientist could predict the likelihood of slope failure during the construction of road cuttings from a knowledge of the linkages between material properties, moisture regime, and slope angle. This type of model may rely upon design equations which necessarily embrace relatively few variables.

Statistical models involve attempts to derive functional relationships between individual components of the conceptual model being considered. The Universal Soil Loss Equation detailed in Chapter 12 provides an example of how it is possible to produce quantitative estimates of soil loss from a knowledge of topography, soil type, rainfall conditions, and land-management practice. This technique could in turn provide a basis for predicting the impact of different cropping patterns and of conservation practices on soil erosion. Likewise functional relationships between river channel dimensions and measures of runoff and sediment transport could be used to evaluate the potential impact of reservoir construction, river regulation or inter-basin transfer on river channel morphology.

Simulation models move beyond the use of statistically-derived functional relationships and involve attempts to simulate the dynamics of the processes involved, often on a continuous basis. Whereas many statistical models are essentially empirical in nature and relate to the range of conditions from which they were derived, simulation models possess the potential for modelling environmental change from a knowledge of process dynamics and are essentially deterministic. The greatest advances in this sphere of modelling are associated with atmospheric and hydrological processes. Thus Thuillier (1976) described the application of regional atmospheric quality models to the San Francisco Bay area which permit the simulation of spatial variation in the concentration of ozone, nitrogen oxides, hydrocarbons, and carbon monoxide at a regional resolution of 25 km^2, and Koch *et al.* (1976) outline the use of a multiple-source urban diffusion model which has been used to analyse the effectiveness of various SO_2 control strategies in San Francisco and Boston. A useful example of a hydrological simulation model which predicts the movement of nutrients and pesticides from agricultural land into water courses is the Agricultural Runoff Management (ARM) model described by Donigian and Crawford (1976). This model simulates the occurrence of runoff, snow accumulation and melt, sediment loss, pesticide-soil interactions, and soil nutrient transformation on small agricultural watersheds. The lack of availability of detailed data for validation purposes presents a problem with models of this type and emphasizes the need for monitoring programmes. We are, therefore, some way from seeing this and similar models of nonpoint pollution processes being widely used but their potential for improved management of agricultural land is clearly apparent.

Simulation models have also been developed for other suites of landscape processes including cloud seeding projects, coastal stability, soil profile development and land subsidence. In the latter case, Gambolati *et al.* (1973, 1974) have described a mathematical simulation of the subsidence of Venice, and Final and Farouq Ali (1975) have demonstrated the possibility of simulating ground deformation associated with oil field development on the Bolivar coast of Western Venezuela.

A number of examples of the practical application of models to assist in environmental impact assessment are provided by studies of proposals for large-scale development of the shale-oil industry and coal mining in Colorado, Wyoming and Montana, USA. Guy (1977) outlines how various statistical models were used to furnish sediment data relating to the potential environmental impact of an extension of open-cast coal mining near Decker Montana. Information was assembled concerning erosion and sediment yield under natural conditions and after reclamation of the mined area, sedimentation of adjacent reservoirs during the mining period, and problems associated with diverting six ephemeral streams away from the disturbed area during the period of mining activity and returning these streams to a man-made valley across the reclaimed spoil upon completion of mining. More sophisticated simulation techniques are currently being employed to evaluate the effects of increased coal exploitation and associated economic development on the regional water resources of the Yampa River basin in Colorado and Wyoming (Steele, 1978). Projections indicate that coal production in this area could triple in the next fifteen years and the population is expected to double or triple over the same period. Much of the mined coal will be converted to electricity or gas locally. Seven coal-resource development alternatives have been identified for this region and models are being used to evaluate hydrological problems associated with wastewater discharges, with the construction of proposed reservoirs, with increased erosion and sediment yield from surface mining and construction activities, and with disruption of groundwater recharge areas by mining and degradation of groundwater quality from disposal of residuals from coal-conversion processing plants. For example, maps have been produced indicating likely increases in the dissolved solids concentration in groundwater around two residual disposal areas after 20, 60, and 200 years.

Current advances in systems analysis permit the development of comprehensive dynamic models incorporating both environmental and socio-economic sub-models and numerous interactions and feedbacks. An excellent example of work of this nature is provided by the Obergurgl model (Himanowa, 1975). This work was undertaken in Austria under the auspices of the Man and the Biosphere Programme and with the cooperation of the International Institute for Applied Systems Analysis (IIASA). It involved analysis of human impact on the alpine ecosystem around the village of Obergurgl and elaboration of the options available for future development over the next twenty to forty years. Consideration of recreational demand, population growth, economic development, land use change, erosion, and landscape aesthetics were included in the model. Feedback loops were able to simulate such processes as decline in summer tourism consequent upon degradation of landscape aesthetics by increased recreational pressure.

15·5 THE PROSPECT

Written over 110 years after George Perkin Marsh's *Man and Nature* and following after a century of further work, a decade of papers, and a decade of readings (Chapter 1), this text cannot claim a novel message or revelative conclusions. Nevertheless, it is hoped that the emphasis on processes and the attempt to highlight and elucidate the role of man in influencing the dynamics of the physical landscape provides its justification. Its chapters demonstrate that man must be viewed as a potent influence on contemporary processes, and they provide convincing support for an extension of Ackermann *et al.*'s contention (1971) to encompass a view that the physical geographer who does not recognise the importance of man in influencing environmental processes 'will be left back in the nineteenth century'. Recognition should also generate an awareness of the many environmental problems that may result from these influences and encourage the physical geographer to play a role in generating increased understanding of the potential impact of future activities. Incorporation of knowledge of the various interactions between human activity and physical landscape processes into models, should provide a means of extending the scope of environmental impact assessments beyond enumeration of general areas of impact towards quantitative estimates of the nature and extent of that change.

The present provides evidence of considerable change within the physical environment as a consequence of the impact of human activity. One can only speculate on the extent to which the pace of change will quicken as a result of technological advancement and increasing population pressure. It could be argued that man's power to disturb is even now running ahead of methods available for assessing and responding to environmental impact, and the theme of this book must continue to represent a profitable avenue for future study and research.

REFERENCES

ACKERMANN, W. C., WHITE, G. F. and WORTHINGTON, E. B. (eds.), 1971, *International Symposium on Man-made Lakes.* Amer. Geophys. Union, Geophys. Monog., 17.

COBB, E. D. and BIESECKER, J. E., 1971, 'The national hydrologic bench-mark network.' *US Geol. Survey Circular,* 460–D.

DONIGIAN, A. S. and CRAWFORD, A. H., 1976, 'Modelling pesticides and nutrients on agricultural lands.' *US Env. Protection Agency, Env. Protection Tech. Ser.,* Dept. EPA–600/2–76–043.

FINAL, A. and FAROUQ ALI, S. M., 1975, 'Numerical simulation of oil production with simultaneous ground subsidence.' *Jnl. Soc. Petroleum Eng.,* 15, pp. 411–24.

GAMBOLATI, G. and FREEZE, R. A., 1973, 'Mathematical simulation of the subsidence of Venice, 1: Theory' *Water Resources Res.,* 9, pp. 721–33.

GAMBOLATI, G., GATTO, P. and FREEZE, R. A., 1974, ' Mathematical simulation of the subsidence of Venice, 2: Results.' *Water Resources Res.,* 10, pp. 563–77.

GUY, H. P., 1977, 'Sediment information for an environmental impact statement regarding a surface coal mine, western United States', *Erosion and Solid Matter Transport in Inland Waters,* IAHS Publ., 122, pp. 98–108.

HIMANOWA, B., 1975, 'The Obergurgl model.' *Nature and Resources,* 11, pp. 10–21.

KOCH, R. C., FELTON, D. J., and HWANG, P. H., 1976, 'Sampled chronological input model (SCIM) applied to air quality planning in two large metropolitan areas', *Environmental Modelling and Simulation, US Env. Protection Agency Publ.,* EPA–600/9–76–016, pp. 92–6.

LEOPOLD, L. B., CLARK, F. E., HANSHAW, B. B. and BALSLEY, J. R., 1971, 'A procedure for evaluating environmental impact.' *US Geol. Surv. Circular,* 645.

NATIONAL ACADEMY OF SCIENCES, 1976, 'Implementation of the global environment monitoring system.' *Int. Env. Programs Committee Dept.*

NELSON, J. G. and SCACE, R. C. (eds.), 1974, *Impact of Technology on Environment: Some Global Examples* (University of Calgary).

ROCKEFELLER FOUNDATION, 1977, *International Problems in Environmental Monitoring* (Rockefeller Foundation, New York).

SCOPE (SCIENTIFIC COMMITTEE ON PROBLEMS OF THE ENVIRONMENT), 1973, *Global Environmental Monitoring System: Action Plan for Phase 1* (SCOPE, Paris).

1975, *Environmental Impact Assessement: Principles and Procedures* (SCOPE, Paris).

STEELE, T. D., 1978, 'Assessment techniques for modelling water quality in a river basin affected by coal-resource development.' Paper presented at International Symposium on Modelling the Water Quality of the Hydrological Cycle, IAHS/IIASA, Baden, Austria, September, 1978.

THUILLIER, R. H., 1976, 'Air quality modelling – a user's viewpoint', *Environmental Modelling and Simulation, US Env. Protection Agency Publ.,* EPA–600/9–76–016, pp. 35–9.

WHITE, G. F. (ed.), 1977, *Environmental Effects of Complex River Developments* (Westview Press, Boulder).

WORTHINGTON, E. B. (ed.), 1977, *Arid Land Irrigation in Developing Countries: Environmental Problems and Effects* (Pergamon, Oxford).

Index

Subjects indexed below are restricted to major references and locations and authors are not included. References to illustrations and tables are indicated by numbers in italics. Mr B. Gomez kindly compiled the index.